When the Sun Darkens

Orbital History and 2040 AD

Return of Planet Phoenix

by Jason M. Breshears

The Book Tree
San Diego, California

When the Sun Darkens
© 2009
by Jason M. Breshears

ISBN 978-158509-117-1

Editor: Paul Tice
Cover Art by Jason M. Breshears

Printed in USA on Acid-Free Paper

Published by
The Book Tree
P O Box 16476
San Diego, CA 92176
www.thebooktree.com

We provide fascinating and educational products to help awaken the public to new ideas and
information that would not be available otherwise.
Call 1 (800) 700-8733 for our FREE BOOK TREE CATALOG.

Acknowledgements

A special thanks to three individuals instrumental in the completion of this work. The author's father, Dan, has spent a small fortune providing the author with hundreds of contemporary and historic texts that are cited thoughout this and the author's prior work, *Lost Scriptures of Giza*. Paul Tice of The Book Tree has provided valuable insight and made possible the publication of this work. We are further grateful to Jane Eichwald for her work in editing and manuscript preparation.

Special Note: Throughout the text the author references a number of his other works, some of which are yet to be published at the time this work was released. If by chance a desired book cannot be found, please contact The Book Tree for more information.

And I beheld when He had opened the Sixth Seal, and, lo, there was a great earthquake; and the sun became black as sackcloth of hair, and the moon became as blood. And the stars of heaven fell unto the earth. . . and the heaven departed as a scroll when it is rolled together; and every mountain and island were moved out of their places. . .

—Sixth Seal of *Revelation*
6:12-14

Table of Contents

I.   Existence of Planet Phoenix 7

II. Age of the Phoenix and Cycle of Cataclysm 13

      Sun Darkening of 583 BC 13

      Sun Darkening of 1135 BC 15

      Sun Darkening of 1687 BC 16

      Sun Darkening of 2239 BC 20

      Chart: Age of the Phoenix 21

III. A Planet of Two Calendars 25

      Chart: Calendrical Synchronicity of Age of the Phoenix 29

IV. Phoenix Cycles Demonstrated 33

V.   Cursed Earth Periods 41

VI. Modern Cursed Earth Periods 51

VII. Orbital Chronology of Planet Phoenix 61

      Chart: Orbital Chronology of Planet Phoenix 73

VIII. Secret Calendar of the Great Seal of the United States of America .. 75

      Chart: Giza Course Countdown Calendar 77

      Chart: Cursed Earth Chronology 80

IX. 2040 AD Return of Planet Phoenix 85

X.   The Joshua Comet Group 93

XI. Vials of Phoenix Comet Group 101

Conclusion of the Phoenix Thesis 107

Appendix A: Effect of 2040 AD Phoenix Transit 109

Appendix B: Legends and Myths of the Sun Darkening 111

Appendix C: When the Sun Stood Still 113

Bibliography of Cited Works 115

Notes and References 119

About the Author 126

I

Existence of Planet Phoenix

Absolutely nothing is being added to the sum of knowledge as a result of original research; indeed not even the discoveries made by people long ago are thoroughly assimilated.

—Pliny, *Natural History*

The Legend of the Phoenix is of particular antiquity. The old stories of the Phoenix held that it began as a great bird in the heavens that immersed into the sun with red plumage, but that it underwent a transformation into ". . .a flaming, leaping *star*." (1) Zecharia Sitchin provides evidence that the ancients associated Phoenix to a great spherical object that appeared on the horizon, (2) a mysterious celestial creature that appeared only every 500 or so years. The legend was universal. It has now been determined that every Native American culture from Alaska to Tierro del Fuego preserved stories concerning a gigantic bird so huge it darkened the sun. (3)

The concept of the Phoenix seems to be a Greek preservation of Egyptian scientific records. A leading authority of Greek studies, Robert Graves, wrote that Phoenix is the masculine form of Phoenissa, a title for the moon meaning *the red, or bloody one.* (4) This fact is an astronomical relic preserved in the garb of fable like the red and golden plumage of the Phoenix, for as we will learn in this work, when planet Phoenix transits between the Earth and the sun it casts a *dark red shadow* over the moon. The link between Phoenix and Phoenicia is quite apparent. Chroniclers long ago, like Horapollo, used the terms "Phoenix" and "Phoenician" interchangeably for someone who returned from a *long journey.* (5)

The earliest stories of Phoenix are associated with the person and reign of Enoch, (6) who was a prophet, astronomer, king and then emperor. He was in fact the first to reign over an empire (7) and the title Phoenix is the Greek form of the Egyptian Pa-Hanok, or *House of Enoch.* (8) The historian Frederick Haberman notes that the Phoenician chronicler Sanchoniathon, writing about 1000 BC, recorded that Phoenix was once a *man.* (9) The connection between this enigmatic phenomenon and this preflood *chronologist* will be made more

7

apparent later in this book. It is sufficient to understand at this point that the appearance of Phoenix to our predecessors had everything to do with a change of *kingship* instigated by the darkening of the solar orb by some vast object.

Long ago the authority of a king was made known by his *seal*. A tablet bearing an official message from the throne was stamped with the seal of the king. Ancient seals were adorned with images relative to the importance of their owners. Scholars of Near East studies have noted that at various points in history among the Sumerian, Akkadian, Babylonian and Assyrian dynasties the kings of Mesopotamia employed the royal seal of the Winged Disk. Sitchin, in his fervor to provide evidence of the Anunnaki planet of NIBIRU, assumes throughout his works that depictions of this Winged Disk concern this Anunnaki planet. In many early depictions of our solar system found among Mesopotamian ruins the winged planet is prominently displayed as a member of this system, and we are not to confuse planet Phoenix with the real and extremely distance planet NIBIRU, brought to the world's attention by Sitchin. Planet Phoenix had its own special representations, always depicted with *wings*.

It was the artistic Mesopotamian reliefs and seals that gave rise to the belief by later generations that Phoenix was an immense celestial bird, so colossal that when it immersed into the sun (represented by a disk) the solar orb would go dark. The Phoenix bird was thought to resurrect because after a protracted period of time the whole sun-darkening episode would reoccur. The artists depicted Phoenix disks with wings because this was a way to convey that unlike the other planets of this solar system, planet Phoenix *flew a different course* than those common wanderers of the ecliptic plane.

Before we continue it is important that the reader not make the mistake of assuming that this book is about NIBIRU, masked by another name. The NIBIRU that Sitchin writes about is a gigantic planet, the origin of the Anunnaki, with an orbital chronology very different than planet Phoenix. Sitchin does not give us a definitive timeline for this alien planet, but for those interested in reviewing it, see my other work entitled *Anunnaki Homeworld: Orbital History and 2046 AD Return of Planet NIBIRU*.

The erudite and meticulous historian Pliny the Elder truly believed that one can know a lot about something and still not quite understand it. His epic work called *Natural History* is today internationally esteemed, his source materials including 147 Roman writers and 327 foreign authors. He was known to write down excerpts from all the books he studied and his personal philosophy was that every book had something that he could benefit from. His penetrating research led him to conclude that throughout the span of human history there transpired ". . . ominous and *long drawn-out* eclipses of the sun," and that "their cause was hidden by the rarity of their occurrence, and for this reason they are not understood." Pliny added that these strange darkenings of the sun were quite natural and that they transpired at *fixed times*. (10)

Today historians and scientific antiquarians gloss over these references, claiming that Pliny merely discoursed about regular *eclipses*, the transit of the moon between earth and the sun. But Pliny does not mention that the moon is what darkens the sun, a significant fact that such a penetrating researcher would not have omitted. In his work he dedicates time to eclipses and the moon, but sets apart these other darkenings as the effect of something different. As it will be demonstrated herein, eclipses by the moon have nothing to do with what the ancients recorded. Interestingly, Pliny foreknew that his findings would be disbelieved, basing his observation on the predictability of human behavior. He wrote that the majestic things of nature lack credibility because men see things piecemeal, refusing to put the entire picture together using the pieces available to them. (11)

Though Pliny could foresee the future disbelief of critics, he could not have foreseen that his own death would be connected in the most astonishing way to these long drawn-out eclipses he wrote about. In 79 AD Mount Vesuvius erupted violently and buried the Roman luxury cities of Herculaneum and Pompeii under several feet of ash and pumice. During the chaos Pliny received notice that a female friend was in distress and required his assistance. He drew near to the volcano in his rescue attempt and probably from an acute inquisitive mind. Sadly, he was never seen again. The Roman historian died in the area of Herculaneum and Pompeii by the wrath of Vesuvius, never suspecting that directly below these Roman cities were the entombed remains of an even more ancient civilization, cities that had themselves been buried alive by Vesuvius in 1687 BC, the *exact* year planet Phoenix transited, darkening the sun and initiating global quakes and volcanism as we will discover.

For his time, Pliny represented the epitome of human knowledge of the world and the universe. Like Aristotle before him, he believed that all learning proceeds entirely or partially by things that are previously known. But there was a man even before Pliny, who in all the ways that make a sage, was a man before his time. His name was Lucretius and his work is preserved in the 7400 line poem entitled *On the Nature of the Universe*. Lucretius died in 54 BC, over half a century before Pliny was born. The year of his death in 54 BC was also noted for a peculiar rain of spongy iron that afflicted Lucania. (12)

Lucretius adhered to the philosophy that the unseen things of the world could be discovered by studying what was seen. He is famous today for his theory that the entire universe is composed of infinitely tiny *atoms*, a fact now of course proven. Lucretius was troubled by the assertions of some that the world was very old because the whole sum of Greek knowledge of the past merely spanned back to the Theban and Trojan Wars (1244 and 1229 BC). He further wrote that the history of the world was one of *earthquakes* that caused entire cities with their citizens to sink to the bottom of the seas, and that the Great Flood that destroyed whole races of men was only one cataclysm in the cycle of quakes. (13) Lucretius' most incredible statement is that *some unknown body*, not the moon, passes over the surface of the sun and *darkens the earth* at a *fixed time*. (14)

Lucretius' writings are more specific than Pliny's, the former detailing the effect and the latter explaining the cause of the sun darkenings, that something passes over the surface of the sun. Lucretius wrote that an *object* passed between earth and the sun while Pliny mentions that these were long *drawn-out* eclipses. While Pliny only infers that this phenomenon was not caused by the moon, Lucretius is more direct. Scholars today vehemently assert that references like these always refer to our moon. Their position is untenable. This explanation is held, however, because of the obvious fact that in the known solar system only two planets transit between the sun and earth, these being Mercury and Venus. Both of these through a telescope are viewed against the backdrop of the glowing sun as tiny dots passing over the surface of the fiery sphere. But this assertion presupposes that the object darkening the sun travels along the ecliptic with the other known planets. As the pseudo-mythical evidence has already revealed, planet Phoenix is flying a *different path* (off the ecliptic plane) and travels far away from the inner solar system.

The hypocrisy of the modern scientific world is abundantly evident in their prejudice against the statements of these men, Lucretius and Pliny. While these philosopher-historians are accredited today with the discovery of so many modern concepts and ideas, they are also accused of not comprehending the simple mechanics of an eclipse caused by the moon. This hypocrisy is amplified by the curious fact that it is largely held by the establishment that the phenomenon of eclipses was actually discovered over 500 years before Lucretius or Pliny were even born. Thales of Miletus was accredited with this, and his story is another strange coincidence in the history of planet Phoenix.

The renowned Father of History, Herodotus of Halicarnassus in his *Histories* wrote that Thales of Miletus was a Phoenician by remote descent. He is named among the Seven Wise Men of Greece. Thales was the first among the Greeks to travel to Egypt and receive scientific instruction. Hieronymous of Rhodes said that Thales was an independent researcher having never had a teacher until he studied in Egypt and associated with the priestly colleges. (15) Thales lived from 640-546 BC. On his return from Egypt he founded the Ionian School of Astronomy and Philosophy. He is thought to be the originator of algebra and he taught that the stars were *solids*, as well as the sun and moon (as we now know). Thales knew that the moons' light was borrowed from the sun, by reflection, comprehending that the moon disappeared when the earth cut off the sun's light (lunar eclipse). He was further aware of the sphericity of the earth. (16)

There can be no doubt that Thales was an initiate to a secret corpus of astronomical information concerning planet Phoenix. He knew more than Lucretius or Pliny, for Thales was in possession of the knowledge of this mysterious planet's actual *orbital history*. In the 4th year of the 48th Olympiad, Thales predicted that the sun would darken for a period of time after the passage of two years. His prediction was made in 585 BC. (17) He knew this would transpire in what translates into our modern calendar as 583 BC, in the reign of Alyattes, King of Lydia, but he merely foretold the *year*, not the day. This fact alone proves that Thales predicted something unusual, not an eclipse. Again, this is glossed over by the thinkers of our time. But what cannot be ignored is what occurred in 583 BC. . . in accord with a specifically *fixed* time.

II

Age of the Phoenix and Cycle of Cataclysm

*Time is measured not by the calendar but by
the events that occupy it.*

—Lewis Mumford,
Technics and Civilization

Thales lived along the western coast of Asia Minor in Ionia not far from the ruins of Troy. The empire of his day was Lydia and by 583 BC King Alyattes of Lydia had been engaged in a war against the Medes of the east, at the eastern edge of Mesopotamia. This date is exactly three years after King Nebuchadnezzar II of Babylon destroyed Jerusalem and the Temple in the days of Daniel the prophet and Jeremiah. Persia was growing and would soon overcome the might of Babylon once allied to Media. This is the backdrop behind this most unusual event.

Sun Darkening of 583 BC

Herodotus had traveled throughout the ancient world of Greece, the Aegean, Asia Minor, Syria, Babylonia, Persia, Phoenicia and Egypt, collecting accounts about everything interesting that different peoples believed. He was meticulous and his discoveries in the histories of various nations were recorded for all posterity in his world-famous *Histories*, an extensive tome now over twenty-four centuries old. He wrote, "My business is to record what people say, but I am by no means bound to believe it." (1) With the casual grace of a patient historian, Herodotus set out the following facts:

> ". . .war subsequently broke out between the two countries and continued for five years, during which both Lydians and Medes won a number of victories. One battle was fought at night. But then, after five years of indecisive warfare, a battle took place in which the armies had already engaged when *day was suddenly turned to night*. This change from daylight to darkness had been foretold to the Ionians by Thales of Miletus, who fixed the date for it in the year in which it did, in fact, take place. Both Lydians and

13

Medes broke off the engagement when they saw this darkening of the day; for they were more anxious than they had been to conclude peace, and a reconciliation was brought about by Sennesis of Cilicia. . ." (2)

Herodotus mentions *three* times in this abbreviated passage that the sun darkened, and not once did he refer to the moon. As eclipses in those days were known, but not with predictive value, it was equally known that it was the moon that caused very short-lived momentary darknesses. Regular eclipses were recorded by Herodotus and in these episodes the learned historian clearly mentioned the *moon*. As Xerxes of Persia marched toward Greece as recorded by Herodotus, the moon eclipses the sun for a short time. The Persians welcomed this omen as portending their victory over Greekdom, which venerated the sun as opposed to the lunar-veneration of Persia and the eastern nations. This occurred just prior to the Battle of Salamais, where, in contradiction to their interpretation of the eclipse, they were defeated by the Hellenic League Greek city-states, led by the warlike Spartans in that year of 481 BC. Herodotus clearly perceived the difference from the 481 BC eclipse involving the moon and the sun darkening event that stopped a raging battle in mid-career in 583 BC. Both Thales and Herodotus have to their credit the support of the unusually intelligent *and* critical thinker Aristotle, who acknowledges both men as his scientific predecessors. (3)

Now, our focus is on the astronomical facts underlying the Legend of the Phoenix. At a *fixed time* according to Pliny the sun darkens, and Lucretius asserts that this is caused by *some other body* that so infrequently appears that its rediscovery in later generations is impeded by the lapse of time. As the Euphratean records, reliefs and seals depict, Phoenix is a winged planet. Therefore, we are to search for something off the ecliptic plane, a planet that flies in and out of our system on an entirely different path around the sun. The ancient traditions claim that Phoenix appeared and darkened the solar orb only every 540 to 550 years, though some sources cite that the period was about 500 years. As the legends are guardians of latent astronomical facts, we must look deeper into the past in our search to unveil this mystery.

Because Thales predicted the exact year the sun would darken, we must assume that he had access to historical records concerning this next, and more ancient, solaric anomaly.

Sun Darkening of 1135 BC

This amazing year exemplifies how the past forms a predicate for the future. This date is 552 years before the 583 BC darkening of the sun during the reign of Nebuchadnezzar II of Babylon. But now, in 1135 BC

Nebuchadnezzar I ruled Babylon, and, discovered among his old stone tablet annals is a record concerning how, in this year, a *comet darkened the sun*. (4) This fact demonstrates that the Babylonians were *aware* of the timing of the return of Phoenix, which they believed was a comet, for they named their king "Nebuchadnezzar II," knowing that the sun-darkening of his centuries-old predecessor, Nebuchadnezzar I, was about to transpire again. And of course we have seen that in 583 BC the sun did darken. The concept of *kingship* is made more profound by the historical fact that in this year of 1135 BC Tiglath-Pileser I became king of Assyria and began his famous military campaigns, subduing all of Mesopotamia. At this time the Assyrian monarchs employed as their Great Seal the *Winged Disk*.

Historical astronomers have also found that at this date the Chinese left records that the sun darkened. They would have viewed this from an angle, for the same event is preserved in the Irish traditions concerning the Tuatha de Danaan. These people, migrating from continental Europe, campaigned against the prior residents of Ireland and Britain called the Fomorii, a race of giants descended from Ham. (5) The two armies of these people met in the Field of Towers and committed to war in the Battle of Magh-Tureadh. Just as had happened in 583 BC, now, in 1135 BC, during the battle, the *sun darkened*. (6) This fierce war is recorded in the *Annals of Clonmacnois*.

The Danaan were of mixed ancestry, Semitic Achaeans that fought against the Trojans at Troy for King Agamemnon of Mycenae. As at Troy, the Danaan engaged the enemy Fomorii of Ireland in opposing camps separated by a dual between two champions that would fight in single combat. (7) This was the Heroic Code, exactly how Israel would face the Philistines in 1027 BC when David slew the giant Goliath, the Philistine champion. The Philistines were exceptionally tall, shown as much taller than the Egyptians on temple reliefs at Karnak, and were descended from *Ham*. (8) The Danaan of Achaea were originally Greeks who integrated culturally with the migrating maritime Israelite tribe of Dan that departed Canaan after having failed to defeat the Rephaim and Anakim giants in their allotted portions of the inherited lands of Israel. Interestingly, the geographical regions that were to be Dan's inheritance were occupied by the *Philistine* cities of Ashdod, Askelon and Joppa. (9)

Phoenix is described in the Babylonian records as a comet. This is not an inaccurate assessment. However, as it will be seen, Phoenix is larger than any known comet. Having a frozen exterior surface that begins burning off into a long tail millions of miles long ever pointing away from the sun, a comet is merely a tiny object in comparison with the enormous size of planet Phoenix. Their similarity lies with the fact that Phoenix too is encapsulated in immense glacial sheets that begin to burn off into a spectacular cloudlike tail that engulfs the earth, darkening the sun. This is akin to the traditions of Venus, which the early Maya recorded was a planet that exhibited a *tail*. In fact, any moon, planet or fragment encased in frozen liquid seas or oceans that approaches too closely to the sun's surface will give off a brilliant tail. But before we speculate any more on the nature of Phoenix, let us peer even further back into the mists of time in our search of an event 552 years before the sun darkening of 1135 BC.

Sun Darkening of 1687 BC

For over a hundred years prior to this distant date there was an expansive human population explosion throughout the world. Cities made of stone were erected everywhere, all with sacred temple precincts and most of them were oriented around pyramid structures as the nations of the world attempted to replicate the Great Pyramid standing at Giza in Egypt, as this author's prior work, *Lost Scriptures of Giza*, serves to show. Also provided in this book is the history of an extremely old and interesting book known as the *Book of Jasher*, a text often confused with its forgeries. The *Jasher* text is twice mentioned in the Old Testament scriptures as a credible source of historical information concerning the world in patriarchal times.

Within this ancient Hebraic record we learn why the sons of Jacob killed all the men of Shechem as detailed in the Genesis narrative. The king of Shechem, a Canaanite, took by force the sister of Jacob's sons, Dinah. Her indignant brothers slew Shechem and the news spread rapidly throughout Canaan, which at that time was filled with Amorite garrisons from the kingdom of Mitanni, the Amorite capital being Mari. Biblical chronologist Stephen Jones in his *Secrets of Time*, along with some other historians, date the rape of Dinah using the Jasher and other Hebraic writings at the year 1687 BC. It was in this year, within weeks after the sons of Jacob slaughtered the men of Shechem, that the kings of Canaan with Amorite troops assembled for war against the house of Jacob, a mighty host of 13 kings with a force divided into seven hosts led by the Amorite king of gigantic stature named Jashub of the city of Tapnach. Following him were many of those descended from the Giants. As Jacob's sons and their trained servants, 112 men in all, prepared their armor and weapons of war, the patriarch Jacob despaired and cried into the sky asking for God's deliverance. . .

> And when Jacob ceased praying to the Lord the earth *shook* from its place, and the *sun darkened*, and all these kings were terrified and a great consternation seized them. And the Lord hearkened to the prayer of Jacob, and the Lord impressed the hearts of all the kings and their hosts with the terror and awe of the sons of Jacob. . ." (10)

This was, in fact, a global cataclysm that ended several thriving civilizations. Many of the mounds throughout Israel and Jordan today cover burned strata that date to this period. The unusual megalithic cities that baffle archaeologists today because of their most unbelievable locales were destroyed in these worldwide quakes.

The earliest megalithic layers of Crete and Mycenae were formed from the collapse of their once-mighty structures and archaeologists assert that these civilizations met with disaster around 1645 BC. (11) In a span of 37 centuries the 42-year variance between this approximate date and the actual 1687 BC year is a virtual bullseye. Strewn throughout the ruins where the builders dropped them have been discovered tools as if the construction work was abandoned and men fled in haste. (12) This would support the occurrence of an earthquake. One fascinating discovery underneath the ruins of Crete at Knossus was the gruesome find at the mountain shrine of Amenospelia where archaeologists discovered a collapsed temple, wherein the skeleton of a youth was found still bound to a shrine where he was about to be sacrificed. A bronze dagger for the deed was laying beside him. The temple ruins were preserved because the mountain collapsed upon it. This sensational find was made in 1981 AD. (13)

Several of the world's coastlines were permanently changed and entire regions thickly populated vanished under the waves. Among them was the famous Canaanite city of Joppa, washed out by a tsunami. In South America the ancient port-city of Tiahuanaco was thrust 12,000 feet into the sky as the Andes mountains were pushed upward, altering the entire American Pacific-Atlantic coastlines and causing to be abandoned other cities such as Machu Piccu, which was now too cold and high to adequately support crops or comfortable living conditions. When the Spanish explored the ruins of Tiahuanaco they were told by locals that it was a major city constructed after the Flood by giants who disregarded a *prophecy of the coming of the sun* which led to their demise. (14) This "prophecy" was probably none other than the astronomical knowledge of the return of Phoenix to darken the sun and destroy parts of earth that was to transpire in 1687 BC. The ruins are littered with monoliths weighing from 60 to 200 tons that had been riveted together with silver bolts removed by the Spanish. Tiahuanaco's artifacts have been radiocarbon dated at approximately 1700 BC, (15) which is only 13 years off. Lending credence to the traditions of gigantic men at Tiahunaco is the discovery by archaeologists of enormous human skulls at the site that had once belonged to colossal men that stood *nine to ten feet tall*. (16) This corroborates the biblical account of the height of Goliath the giant, who stood 9'9" tall.

Evidence of cataclysm near Tiahuanaco is found north of the city in the enormous ruins of Puma Punka, where 27 ft. long andesite blocks weighing as much as 300 tons were strewn about by some tremendous force. Bolivia's National Archeological Institute has dated the massive ruins at 1700 BC. (17)   West of Tiahuanaco have been discovered ". . .greater monuments, such as large gateways with hinges, platforms and porches made of a single stone." Also, ". . .enormous gateways, made of great masses of stone, some of which were thirty feet long, and fifteen feet high, and six feet thick." (18) The megalithic doors with jambs and lintels all hewn cleverly from a single block are characteristic of those same peculiar rock doors found by Porter north of Israel in the region of Bashan, now located in modern Syria and Lebanon, a territory anciently belonging to the Rephaim giants, famous in the Old Testament.

Other sites dated at this time that have the same unusual polygonal and megalithic masonry are the ruins of Palenque, Cuzco, Kalasasaya, the Fortress of Sacsayhuacan as well as those in the jungle depths of the Yucatan. Just a day's walk from the ruins of Bashan lies a truly gargantuan cut and dressed stone still lying in situ in its quarry at Baalbek in Lebanon, a single stone 69 feet in length. Three more enormous stones, called the trilithons, were raised to a high position and set in the walls of a foundation for some unusually large platform, said in the earliest records of the site to have been the location of a pre-flood city. Archaeologists have noted that the project seemed to have come to an abrupt halt, for the craftsmen's tools were found lying about the site and the building remained unfinished. It appears that the site was being rebuilt after it had been in a ruinous condition for a long time. As Baalbek parallels Tiahuanaco, this Andean anomaly also mirrors the ruins far off the coast of Peru in the South Pacific, known simply as Easter Island.

Easter Island was once the center of a mighty island empire controlling a vast archipelago of island kingdoms until the ocean level raised considerably, or the sea floor lowered, plunging countless islands beneath the waves. All over Easter Island are half-buried statutes wearing unusual hats and mare (platform pyramids) along with what archaeologists have found to be most puzzling: a heroic statue *70 feet long still lying in the quarry.* (19)   Like the Baalbek monolith, this statute is still attached to natural rock. Further mirroring Baalbek, at the Easter Island areas where the giant heads were being erected, archaeologists have found the craftsmen's tools still lying about where *they were dropped* when all construction ceased. Entire workshops strewn with tools that were never touched again. (20)   The islanders claim that ever since the disaster, traveling overseas was discontinued. (21)   Far north of Easter Island in the Pacific on Ponape Island between Japan and Hawaii are the basalt ruins of the city of Metalanim. This great city is 11 square miles of earthworks and megalithic masonry, colossal pillars, temple edifices and artificial waterways of dense basement stone that could have maintained a population of two million people. The city was destroyed by quakes and flooding and an evident alteration of its coasts. For a long period of time Metalanim was underwater. We have no written records or traditions concerning it. The same fate at the same time ended the civilization on Yap Island in the Carolinas.

This catastrophe did not pass by China unnoticed. A series of yellow fogs, a famine and the *darkening of the sun* (22) revealed the Mandate of Heaven that kingship in China must be changed, and this disaster brought an end to the ruling Chieh Dynasty as the T'ang Dynasty emerged. In the year 279 BC, grave robbers in the Honan Province of China excavated a royal tomb and discovered what has come to be called the Bamboo Books. In these texts we learn that a terrible cataclysmic episode occurred in 1688 BC (one year variance) of terrible earthquakes and strange astronomical omens as well as a rain of meteoric rock. (23) This dating and disaster is confirmed in Chinese *seals*. A fragment of the Bamboo Books reads:

> XVII. "The Emperor Kwei. In his tenth year, and in Shin, 29th of the cycle, the five planets went out of their courses, in the night the stars fell like rain. The earth *shook*, the J and the Joh rivers (in Honan) became dry." (24)

Asia provides us with more detailed data on this event. The cultures of the Gobi and Takla Makan deserts of China were once thriving forests of people and trees but the climate of the region was drastically altered. This may be related to an observation made by the prolific researcher David Hatcher Childress who wrote that the Mongols called the Gobi desert by the name of *Shamo,* which he believes is akin to the Middle Eastern deity named Shamos, who was worshipped as a *black star* that was considered an evil heavenly body. (25) Planet Phoenix. Also in Cambodia lies the formerly lost city of Angkor Wat, another city where archaeologists excavating the earlier parts of the ruins found that it was littered with stone tools that were left behind. (26) Angkor Wat was abandoned, reoccupied, deserted and discovered again several times in its long history.

Pliny the Elder died during the volcanic activity of Vesuvius in 79 AD. At this time the Roman luxury-cities of Herculaneum and Pompeii were buried and archaeologists have excavated marvel after marvel of preserved human bodies, animals and priceless artifacts as well as frescoes and murals. But these scientists have found something even more profound. Unbeknownst to the Romans, Herculaneum and Pompeii were built directly over more archaic settlements that had been themselves buried under the wrath of Vesuvius. Geologists have found that this prior eruption buried everything 90 miles from the volcano and testing has set this earlier volcanic entombment at approximately 1780 BC. (27) This is merely a professional estimate, albeit a good one. It is only 93 years off from 1687 BC.

Stonehenge on Salisbury Plain in Wiltshire, England is one of the most misunderstood monuments in the world. Sir Norman Lockyer studied the site and concluded after extensive research that the Stonehenge complex was erected in *1680 BC*. (28) His dating is of course seven years off, however, he merely discovered the date it was toppled and *renovated*, being Stonehenge III. Perhaps not a coincidence, this intriguing fact was published on pages 5 and 52 of David Davidson's monumental book, *The Great Pyramid: Its Divine Message* (note again, the 552 year cycle).

We have dared to gaze intently into the obscure history of our world in the search for validation of this thesis, and now, we must seek to glimpse an even more remote past. As we arrived to 1687 BC after a 552 year journey from 1135 BC, 552 years after 583 BC, we must now peer backward another 552 years to the most dreadful and globally-remembered catastrophe of all time. . . the Great Flood.

Sun Darkening of 2239 BC

It is imperative that the reader know that this author did not date the Deluge. The 2239 BC date derives from the meticulous chronology of Stephen Jones in his *Secrets of Time*, a timeline incorporating the biblical chronologies, Hebraic extracanonical works and Assyrian astronomical records. We arrive here simply by adding 552 years to 1687 BC.

The Great Deluge nearly wiped out humanity. The mechanism by which the whole world was destroyed was by a comet impact seven days after the transit of Phoenix. Rabbinical records indicate that the Flood was caused by the appearance of *two stars*, (29) one being Phoenix and the other a comet that crashed into Earth, initiating the flooding. The Great Flood is fully detailed in this author's work, *Anunnaki Homeworld*. The comet impacted in the area of the Gulf of Mexico.

The antiquarian and author Frank Joseph in his book *The Destruction of Atlantis* cites several traditions and records concerning the darkening of the sun at the Deluge, (30) but one such text that escaped him is our cherished *Book of Jasher*, and in this early Hebrew historical chronology of the ancient world we read that seven days before the flood –

Chart: Age of the Phoenix

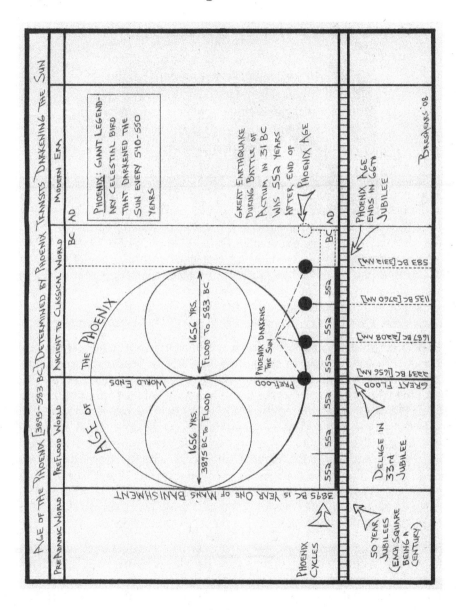

> "...the Lord caused the whole earth to shake,
> and the *sun darkened*, and the foundations
> of the world raged, and the whole earth was
> moved violently, and the lightning flashed,
> and the thunder roared, and all the fountains
> in the earth were broken up, such as was not
> known to the inhabitants before; and God did
> this mighty act, in order to terrify the sons of
> men, that there might not be more evil upon
> the earth... and at the end of seven days... the
> waters of the Flood were upon the Earth." (31)

Planet Phoenix transited, darkening the sun seven days before the Flood commenced. Dr. Frederick Filby, author of *The Flood Reconsidered*, theorized that the Deluge may have been caused by the close pass of a wandering *minor planet* that collapsed the marine canopy once enveloping earth. (32) This theory is supported by the geographer Donald Patton, author of *The Biblical Flood and the Ice Epoch*. He believes that not only did a planet come too close to Earth but this body carried with it comet-like debris that rained on earth and that some of that water contributed to the Flood. (33) The Phoenix appearance in this year explains why the Black Sun has been associated with stories of the Flood. (34)

The Legend of the Phoenix began with the transit of Phoenix between earth and the sun's surface in antiquity when Phoenix was covered in a layer of frozen liquid. The planet's proximity to the sun melted the ice and created a thick tail of gases and rock detritus that fragmented off the surface of the dead planet. The tail, just as with comets, points away from the sun due to solar wind and to the people of the Earth long ago this spectacle appeared as a celestial *sword* approaching the planet, a symbol for judgment from God. When Earth passed into this tail of debris, planet Phoenix was directly between Earth and the sun and the tail in the skies, as seen from the surface of our world, would have appeared as vapory wings enveloping the sun causing the solar light to be choked out to a dull dark brown dullness. The Winged Disk was the seal of those kings that ruled over men during the Age of the Phoenix.

The Phoenix legend died out after 583 BC, some Greek naturalists and writers even assuming ridiculously that the Phoenix was a bird that periodically migrated to Egypt. This was a desperate attempt to account for the cogent fascination of the ancients in the appearance of something they called Phoenix. From a variety of early textual sources from Greek historians, Babylonian annals, Irish traditions in the *Annals of Clonmacnois*, biblical and Hebraic texts, Chinese Bamboo Books and seal inscriptions, as well as input from modern authors and chronologists, we have in these few pages easily reconstructed the astronomical phenomenon underlying the badly misunderstood yet historically persistent Legend of the Phoenix. With a casual walk through the pages of history we now understand how Thales *predicted* the darkening of the sun, why Lucretius believed the sun darkenings were caused by *some unknown body* and why Pliny stressed that these eclipses of the sun were *unusually long*.

By unearthing the past do we see the future, and this tenet is no less true as when acknowledged by Thales. Phoenix still orbits the sun, it has been seen by modern astronomers, it will return at a *fixed time*, it is seen with perfect clarity in the apocalyptic book of *Revelation*, but first we must look even further back through history into the times *before* the Flood. . . even into the pre-Adamic World.

III

A Planet of Two Calendars

After having viewed our present condition, it may be well for us to look back and review our former history, and get a knowledge of the state of the world in former times.

—Jewish Rabbi Hillel III
(358 AD) (1)

We have now discovered four distinct dates in ancient history all linked by three facts. The sun darkened each time, this occult of the sun occurring every 552 years and these events were specifically recorded by the hands of men. The first events were in 2239 BC and the last in 583 BC, and the intervening time between is precisely *1656 years* (552 x 3). This fact alone demonstrates for us that the Phoenix legend and cycle is the key that unlocks the secret to mankind's most archaic and *important* calendar.

In the most distant epoch of humanity there was a year when mankind was *banished* from his terrestrial paradise home and sent out to wander the earth under a curse. The Creator had given men the ability to live free from doubt and constraints, but by listening to an Interpreter (serpent), mankind fell under the curse and power of those beings that defined for them *good and evil*. Once mankind's eyes were opened (sinister intent of the Anunnaki) he was no longer beyond good and evil, but was from then on a slave to his own conscientiousness. Now his conscience his guide, man was cursed to condemn *himself*. He is banished into the world for a period of 6000 years (120 fifty-year Jubilee periods) until the return of the Chief Cornerstone that the Builders (Anunnaki) rejected. . . the Savior.

During the current 6000 years of exile the redeemed among men shall refill the positions in the Government of God emptied by those Anunnaki that rebelled, the Fallen Ones. Once the number of the souls of men completes the spiritual pyramid, then will the Stone that is made Head of the Corner descend and *remove the curse* in the 6000th year.

The Phoenix Cycle of 552 years provides us with the elusive and enigmatic year that begins this 6000 year timeline. The Hebrew year for the Great Flood, derived easily by calculating the birth-death genealogies in the

25

book of Genesis, was the year *1656*. This 1656 year period (552 x 3) is from mankind's banishment from Eden to mankind's destruction by the Flood in 2239 BC, which we already deduced began a 1656-year countdown to the 583 BC sun darkening predicted by Thales. Thus, the latter 1656 years we have covered is merely a reduplication of the 1656 years of the pre-flood world. This gives us the date 3895 BC as the beginning of the 6000 years of man's banishment. This time covers exactly six Phoenix Cycles of 552 years each. Though the Hebrew chronology accurately dates the Flood at 1656, chronologies after the Deluge are horribly obscured and further rabbinical torture of texts and chronologies has resulted with a modern Hebrew Calendar that is 133 years off today.

In Genesis the events of Year One (3895 BC) concern how and why God cursed earth. (2) In Genesis as well as in several other Hebraic writings, we find mention that at this time appeared a "fiery flaming sword." It appeared in the east. The sword is a symbol of judgment, reconfirming to mankind that the curse had begun and would not be lifted until the appointed time. To men the spectacle was of an immense celestial blade in the heavens, but today we recognize that what was viewed was a large comet-like object with a long tail. Phoenix. The comet/rogue planet has now become the SYMBOL FOR JUDGEMENT ON EARTH. This meaning will hold true throughout this book even into the coming Apocalypse.

Priests and scholars in antiquity strived to determine Year One. Their studies developed into what was by Roman times called the *Annus Mundi* system, meaning "Year of the World." Alexandrian scribes and historians in Egypt were familiar with the Annus Mundi chronology and spent their time attempting to correct the disparities found between variant timelines. All too often overzealous priests or Christian authors altered their own sacred calendrical systems to suit their socio-religious purposes and then tried to justify the changes by finding parallels in Annus Mundi dating, while others modified Annus Mundi dates to correspond with assumed calendrical points in other dating methods. Over time this dating system has resurfaced, been lost again, reverse-researched to again reassemble for a while before falling back into obscurity. . . ever the victim of Roman and Muslim invasions resulting in a loss of texts and ruined libraries, fanatical Christian and Islamic book burnings or lost into the private collections of old Europe.

The Greek astronomer and geographer Eratosthenes was in charge of the famous Library of Alexandria under Ptolemy III. He studied many historical records in the third century BC and was very knowledgeable of Annus Mundi dating. He believed that Egyptian dynastic records contained fragments of Genesis history. (3) He even thought the world was spherical and he divided the earth into 60 parts and calculated its circumference to within 200 miles of accuracy (not too bad for a measure of 24,000 miles). Eratosthenes also drew a map of Europe, Britain, Ireland, the Mediterranean world, Arabia, Asia and the Persian Gulf all the way to the Horn of Africa. He even theorized of unknown continents and was an expert mathematician. We cannot lightly ignore that he used Annus Mundi dating.

Flavius Josephus, author of the famed *Antiquities of the Jews*, was acquainted with the Annus Mundi system and even attempted to date Solomon's reign in Annus Mundi years. (4) Josephus had access to many Temple archives. Another early writer was Syncellus, who in about 800 AD preserved the Egyptian *Book of Sothis,* all in Annus Mundi years. Geoffrey of Monmouth, celebrated author of the *History of the Kings of Britain,* also used the Annus Mundi chronology to ascertain that the birth of Christ was 1008 years after the founding of the city of Londonum. Even in texts from secret societies like the Masons we find that official society documents recorded events in history in Annus Mundi dating, like the Inigo Jones Document and Wood Manuscript that both refer to Enoch and the Great Pyramid as well as other events in biblical patriarchal times. The advantage of dating under the Annus Mundi system is clear – with years recorded in this way we have an *unbroken timeline* from Year One to the present, and even through the future to the year 6000, which under our present calendrical scheme is 2106 AD. This year, 2009, is the year 5903 AM.

Year One (3895 BC) has been determined in six very unique and independent ways:

1. Phoenix Cycles of 552 years, beginning with the sun darkening at the Flood, involving four darkenings of the sun to 583 BC (1656 years) are the origin of the pseudo-mythical traditions of Four Ages, or the Four Suns of olden times. Counting these same 1656 years *backward* beyond the Great Deluge of 2239 BC (1656 AM) provides us with the year 3895 BC as Year One Annus Mundi.

2. Modern chronologist Stephen Jones in his work *The Secrets of Time* identified 3895 BC as Year One, having no knowledge of Phoenix Cycles, by simply calculating the periods of biblical genealogies, king-lists, Hebraic records and Assyrian astronomical texts.

3. Apocryphal text called *Book of Adam and Eve II,* composed at the Alexandrian Library from older source materials, reveals that from Jared's 40th year alive there were still 5500 years remaining until the end of the 6000 years of man's banishment. (5) By calculating the periods of the patriarchs in Genesis we find that Jared's 40th year was the year 500 Hebrew Reckoning/Annus Mundi. Though Stephen Jones does not include any of this data in his research, using his

chronology, we find that Jared was 40 in the year 3395 BC. This reveals that Year One is 3895 BC, and the year 6000 is the year 2106 AD.

4. A conceptual pattern revealing that 3895 BC is year one is by reviewing the amazing synchronicity herein:

3895 BC (1 AM) Man banished from Paradise: *Adam* cursed.

1948 year period to –

1948 BC (1948 AM) *Abraham* born at the calendrical nexus between the BC and AM systems.

1948 year period to –

1 AD (*3895* AM) *Anno Domini* (Year of the Lord) calendar begun counting down to return of Chief Cornerstone

1948 year period to –

1948 AD (5842 AM) *Israel* reborn as a sovereign nation to initiate the Last Days to the Apocalypse and Armageddon

5. Abraham's 1948 BC birth confirmed in the nigh-1000 year old Talmudic commentary of Rashi in Leviticus Rabbi 29:1. (6) The fact that his birth also aligns with 1948 Annus Mundi is not without significance. Josephus in *Antiquities* wrote that Abraham was born in the 292nd year after the Flood, which we know transpired in 1656 Hebrew Reckoning/Annus Mundi. As 1948 AM is 292 years after 1656 AM, and 1948 AM is likewise 1948 BC, we clearly find that 3895 BC is Year One. Josephus refers to Abraham's birth in *Antiquities* 1.6.5.

6. The final method of determining Year One would require the reader to review the astonishing charts and material in this author's work entitled *Chronotecture: Lost Science of Prophetic Engineering*, which carefully demonstrates how the geometrical dimensions and arrangement of the Great Pyramid form a gigantic timeline that records 246 famous historical and *future* prophesied events until the year 2106 AD (6000 AM).

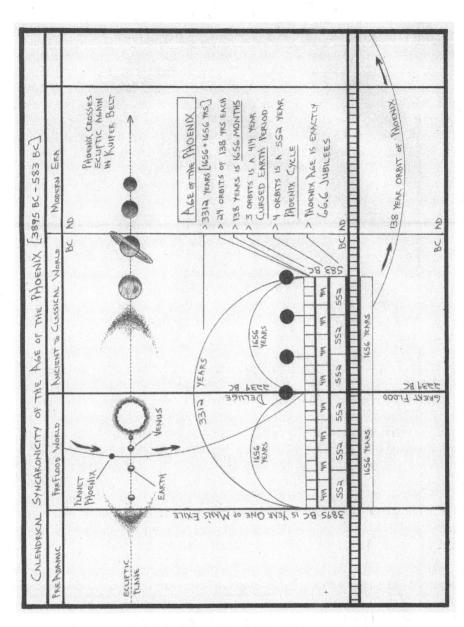

Chart: Calendrical Synchronicity of Age of the Phoenix

The Phoenix Cycle of 552 years seems to end in 583 BC, with the sun darkening during the battle between the Medes and Lydians as predicted by Thales. The Old World was waning and the New World of antiquity we refer to as Classical civilization was one of emerging Greek and Roman states that would eventually result with powerful European and even American nations. Phoenix darkened the sun no more and the legend died out with the cultures that bred it. But this strange visitor to our solar system continued to arrive on time, almost always unseen, and yet its effect was often felt powerfully on earth. Though the orbital synchronicity allowing for direct transits that occulted the sun ceased in 583 BC, planet Phoenix will realign in the future so that the Black Suns of ancient times will terrify the world once more. In fact, it has already begun synchronizing in American history. We will review these instances and modern records later in this work.

The Legend of the Phoenix as transmitted through the Greeks, reveals a venerable advanced astronomical science relative to a formerly known and practiced method of time-keeping. This system was practiced by the Egyptians – evidently the keepers of the secrets – not only of the calendar, but also of the nature and orbital chronology of planet Phoenix. Though Zecharia Sitchin is an undisputed authority on the Anunnaki, and on the existence of NIBIRU, to a fault this scholar automatically links every reference to other planets mentioned by our predecessors that were not among those of the known system to have been NIBIRU. A case in point is in his *When Time Began*, in his chapter Keepers of the Secrets, wherein he notes that on the Papyrus of Queen Nejmet the two Cord Holders measure the angle of a *planet* the Egyptian texts refer to as the Red Eye of Horus. To Sitchin this is a reference to NIBIRU, but more precisely this is Phoenix, which turns the heavens *reddish* and the moon dark red, like blood. Further, the Cord Holding measurements are overseen by a *goddess of the calendar* and her symbol was the stylus of a palm branch, which the Egyptian scribes linked with the "counting of years." The Phoenix connection is easily proven in that the early Greeks, for some unknown reason, referred to date-palms as *phoenixes*, and traditions have, for unknown purposes, always linked the Legend of the Phoenix with these trees. And this is precisely what planet Phoenix allowed the ancient Egyptian chroniclers to do, to *count the years* and maintain the calendar.

To better understand this Egyptian science, we must return to our modern chronologist Stephen Jones. In his research he reveals that there is a definitive period of time noted throughout all of biblical chronology known as a Cursed Earth period involving *414 years* of *curses* which were set upon and later lifted on men or nations. He noted that 1656 years until the Flood was exactly 414 x 4 years. The Flood was a *judgment* introduced by the curse upon mankind and his continued rebellion against his Creator. As 4 is the number for the *Earth* (an arcane concept), four Cursed Earth periods resulted in a global cataclysm because mankind ignored the effects of the curse, rather than acknowledging it and seeking deliverance from his Creator, which would have been provided. What Stephen Jones did *not* know was that the Cursed Earth system of 414 years is entirely *astronomical*, relating totally to the orbital periods of planet Phoenix.

The Cursed Earth system was not unknown in early times. The Alexandrian scholar Aristarchus of Samos was studying at the Library with its half million texts when the ancient *Parian Chronicle* was composed from older sources in 265 BC, (7) a text referring to the Great Deluge. Aristarchus himself had written that he believed that the world was destroyed every 2484 years, (8) a completely arbitrary number that denotes that this man must have had some educated reason for specifying this unusual span of time. We see clearly that 2484 years is exactly *414 x 6*, six Cursed Earth periods. Such a statement infers that he had access to literature 23 centuries ago that we no longer possess. Aristarchus had patience and tenacity, as revealed by the fact that he took the time to draft the first critical editions and commentaries on Homer's epics.

In order to impress upon the mind the reality of planet Phoenix, its 552 year Phoenix Cycle and the 414 year Cursed Earth cycles, this author provides the following chapters which contain timelines of the history of the world interwoven in a matrix of 552 and 414 year periods. These epochs connect events through human history with such amazing synchronicity that we can perfectly conclude the orbital secret to planet Phoenix. . . its actual orbital longevity in *years*. This is simply derived by factoring the differential between 552 and 414. Phoenix enters our inner solar system every *138 years*. This means that a Cursed Earth period was exactly 3 orbits of Phoenix, while a Phoenix Cycle involved 4 orbits.

And this orbital period was specifically known to the ancient Egyptians. Discovered upon Egyptian papyri of the Osirian Texts as exhibited by David Davidson, the Egyptians noted a *1656 month* period of time that was directly related to the Flood. While this could easily be understood as a lunar code referencing a 1656 *year* period (which the Egyptians were fond of doing), this is no doubt a reference to the orbit of planet Phoenix – for 138 years is exactly *1656 months* (138 x 12). This astronomical synchronicity is perfect, further revealing that the Flood was indeed in 1656 Annus Mundi/Hebraic Reckoning.

We have additional evidence of an Egyptian origin of this advanced orbital knowledge of Phoenix's 138 year orbit. It is found in an old Jewish legend. This story explains that Jacob (Israel) in his dream of the Ladder to Heaven encountered two angels that had been *cursed* because they had warned Lot about the impending fiery destruction of the cities of Sodom and Gomorrah. The period of time these two angels were condemned to stay out of heaven was *138 years*. (9)

Before we continue on with our study of this 138 year orbit of planet Phoenix, we shall now delve into the annals of history to see how the very framework of time itself involves Phoenix Cycles and Cursed Earth periods.

IV

Phoenix Cycles Demonstrated

*I have considered the days of old, the years
of ancient times.*

—Psalm 77:5

Our history is fascinating, and virtually unknown by those we trust to teach it. The Phoenix Cycle, according to Mario Reading in his commentaries on the prophecies of Nostradamus ". . . were deemed to herald the beginnings and ends of great human eras." (1) And this is what we have thus far found to be true. We have chronicled the history of the world from 3895 BC to the Flood in 2239 BC, including 1656 additional years to 583 BC when the sun darkened, after being predicted by Thales of Miletus, a total period of 3312 years, or 552 x 6. This was the Age of the Phoenix, the second half of this age having each 552 year period resulting with a disaster and/or a major social rearrangement. This astronomical period involving the orbital cycle of planet Phoenix seems to have set a precedence throughout human history that serves to hint that the future is but an echo of the past.

Though 583 BC appears to end the Phoenix Cycle, this is only because the *effect*, the darkening of the sun, did not reoccur. Further, the time after 583 BC involved a complete paradigm shift in the consciousness of civilization in the Near East and Egypt, as the older Semitic empires fell to Persia, which gave way to the Greeks and later, to the might of Rome. As a result, the Legend of the Phoenix fell into disrepute and obscurity, fallen prey to the imaginative musings of early cryptozoologists. But the cycle remained true, for precisely on schedule Phoenix entered the inner system and back into the annals of history.

Counting 552 years after 583 BC we arrive at the year of 31 BC, when Rome was in a civil war. Marc Antony and the famous Grecian queen of Egypt known as Cleopatra had formed an alliance and their military forces clashed with the powerful Roman navy under the command of Octavian. Antony and Octavian vied for the supremacy of Rome and their armies met and fought in the famous *Battle of Actium*. As the naval war commenced a devastating *earthquake* ravaged the coasts, initially killing 10,000 people from the toppling structures. The final death toll was told to exceed 30,000, as the cities caught fire, burning those unable to free themselves from the

33

rubble. The quake afflicted Judea, with the historian Josephus writing that it was believed to be the worst disaster in the history of Judea. It is an interesting coincidence that Josephus comments on this earthquake in 31 BC in *Antiquities*, in Book 15.5.2 (1000 + 552). Octavian won the struggle, Marc Antony and Cleopatra later committing suicide, with Octavian becoming the famous Augustus Caesar, who initiated a new government throughout all the Roman world – an empire that originally had as its battle standard a picture of the Phoenix.

The quake during the Battle of Actium parallels the quaking of the previous sun darkenings during the battle between Jacob's sons and the Canaanite kings, and also that of the Tuatha de Danaan against the Formorii during the Battle of Magh-Tureadh, as well as the global quakes at the Flood. Planet Phoenix passed through the inner system, but had fallen out of sync with Earth's position around the sun, failing to transit. But the effect of this planetary body's presence still met with the reactionary earthquakes. That these earthquakes were global is indicated by the ruins found of the Olmecs on the other side of the world at the site now called Tres Zapotes. The Olmecs employed the Long-Count calendar before the Maya inherited it, and the last known Olmec date discovered among the ruins of this stone city was found upon Stela C and translates into our modern reckoning as 31 BC. (2) The ruination that happened at Tres Zapotes may have occurred just weeks or months after this last date-inscription was made.

In considering the history of Rome we cannot neglect to review its beginning. The people descended from the Trojans that lost out in the great war in 1229 BC in Asia Minor to a naval invasion of Greek states, led by the Mycenaeans, eventually settled along the banks of the Tiber river in Italy among the Etruscans, Latins and Sabines. They founded a city and called it Rome in 753 BC, which began a 552 year countdown to the end of the Second Punic War in 201 BC when Rome defeated Carthage, a war that began with a *naval invasion* by Hannibal of Carthage against the Roman territories. More on Rome shall be disclosed herein.

In the year 713 BC *every* calendar in the ancient world was changed. A former rock and metal moon of NIBIRU almost collided into earth, known as the Dark Satellite. Its full orbital history is revealed in *Anunnaki Homeworld*. This was the third regal year of King Numa Pompilius of Rome and he, like all other regents of his day, had to modify his calendar from 360 days a year to 365.24 days. The priests of various cultures maintained the 360 day count only in their sacral systems. In 713 BC King Sennecherib of Assyria sought to destroy Jerusalem and the Temple but his army was apparently vaporized in a lightning blast from heaven (a flux tube discharge between Dark Satellite and Earth such as one photographed between comet Shoemaker-Levy 9 and Jupiter in 1994). 552 years after this amazing year was 161 BC, the year the

Jews cleansed the Temple after the passage of the prophesied 2300 days (3) after the Syrian King Antiochus Epiphanes IV desecrated the Temple with pig's blood. Amazingly, Sennecherib is famous in the Scriptures for his letter to King Hezekiah of Jerusalem that claimed that *no* God or gods could deliver Jerusalem from the might of Assyria, and 552 years later Antiochus called himself by the name of *Epiphanes*, which means *god manifest* (is present). This was 161 BC, when a *reddish* rain contaminated ponds and lakes and a plague originating in the Middle East and Persia began afflicting Roman provincial regions. (4)

Additional connections between Assyria and Syria can be found. The emerging eastern power of Media that made war against the Lydians during the 583 BC sun darkening, now, in the year 615 BC under King Artaxerxes, made war against the empire of *Assyria*, which was the superpower of the Middle East at the time. This began a 552 year period to 63 BC when the *Roman* general Pompey conquered *Syria* and annexed Judea as a provincial territory of Rome, the western superpower of the time. Assyria is connected to Rome in another startling way.

In the year 607 BC Nebuchadnezzar II of Babylon (noted primarily in this book for reigning during the 583 BC sun darkening, who was descended from Nebuchadnezzar I who reigned during the 1135 BC sun darkening) utterly defeated the *Assyrian* army. (5) The Ten Tribes of Israel had been forcefully settled amongst the Cimmerian and Scythian peoples by the Assyrians, deported away from Israel in 745 BC, exactly *138 years* before Assyria's army fell to Nebuchadnezzar II of Babylon. Many of these descendants of Israel (now mixed amongst the Germanic peoples) fled Asia and began dispersing throughout Asia Minor, parts of Greekdom and Europe. They had fully assimilated (the Assyrian program) with the Germanic deportees and reappeared at this time on the world scene under their new identities as the Celts and Gauls. These peoples formed kingdoms all around the fringes of the Roman provincial territories throughout Europe as far as Iberia (Spain), which preserves the name Hebrew (Iberia) as does the river Ebro. Nebuchadnezzar's victory began a 552 year countdown to Julius Caesar's Roman invasion of *Britain* in 55 BC, the greatest of *Celtic* strongholds. The Babylonian victory and violence against the formerly oppressive Assyrians was a shock felt throughout the ancient world heralding to the people of the Near East that a change was coming, just as the unprecedented invasion of Britain over the English Channel.

Nebuchadnezzar's victory marked the 666th year since Assyria had first become an empire in 1273 BC, when Assyria annexed *Babylon* and adopted the Winged Disk (Phoenix) as its official Great Seal – an emblem both Rome and then the United States would adopt prior to changing this legendary bird symbol to that of the Eagle. Amazingly, as we will see in Chapter VII, this year of 1273 BC was the exact year planet Phoenix entered the inner system. This was the start of the Assyrian Empire, which would go through phases and later lose territories, only to acquire them again. Such as in the case of the rebellion of Babylon. Exactly 144 years (12 x 12) after 1273 BC the city of Babylon would be reconquered by the Assyrian King Tiglath-Pileser I in 1129 BC. The Assyrian deportation program, removing the Ten Tribes of Israel, lasted 24 years starting in 745 BC. The final year that the remaining tribes were taken into Assyria and settled among other subdued nations was 721 BC under Sargon II, precisely 552 years after 1273 BC (the beginning of Assyrian Empire). Intriguingly, counting 552 years *before* 745 BC we arrive at an earlier invasion of Israel in 1297 BC, when the Canaanite general Sisera subdued the Tribes of Israel and made them pay tribute. They were later delivered by Barak and Deborah in 1277 BC.

In 931 BC (2964 AM) the Ten Tribes of Israel split from the Kingdom of Judah, effecting the establishment of the Two Kingdoms (mirroring the early Egyptian Two Kingdoms) of Israel and Judah. One thousand years exactly after this political subdivision was the fateful year of 70 AD, when the Romans under Vespasian and his son Titus conquered Jerusalem and destroyed the Temple (3964 AM). The Kingdom of Judea was crushed by the Legions and Josephus reported that over a million Jews died defending their city or were executed afterward, and that 97,000 people were sold in the slave markets. The fields were salted (just as the Romans had done to Carthage) and the Temple relics and artifacts were all taken to Rome and reliefs of them were immortalized upon the Arch of Titus. This virtual end of Jewish nationality began a 552 year period to the emergence of another enemy of the Jews: *Islam*, and the start of the Muslim Calendar in *622 AD*, known as the Hijrah. The Hijrah Calendar is adhered to today for the entire Islamic world, a culture that shares an ancestral animosity with the Jews passed down from patriarchal times concerning ownership of the land of Israel/Palestine. Muslim armies invaded modern Israel in their 1948 AD Independence War, Six-Day War and Yom Kippur War.

In 312 AD the Roman realms are again engaged in civil war. On his way to subdue his rival Maxentius, Emperor Constantine the Great sees a comet while en route with his army. The event eventually ignited his conversion to Christianity, or was used as a ploy to justify it. Though we will not indulge in the 414 year Cursed Earth periods until the next chapter, it is not to be missed that this comet appeared in the skies precisely 1242 years (414 x 3) prior to the end of the 6000 year timeline that began in 3895 BC and ends in 2106 AD (6000 AM). This comet was seen over Europe 552 years before the year 864 AD when the Irish Annals record that both a lunar and solar eclipse transpired. (6) While the lunar eclipse may have occurred, astronomers note that no solar eclipses happened in this year. And they are correct, to the extent that the moon did not occult the sun, for we now have seen demonstrated how a *comet* transiting between earth and the sun will darken the solar orb with its debris tail, which envelopes earth. This was not Phoenix, but a large comet with an orbit detailed fully in *Anunnaki Homeworld*.

Julius Caesar was born in 101 BC (3794 AM), a man who brought immense changes to the Roman government and culture. Some believe he is the beginning of the Roman Empire. He became a dictator and was lamented as the cause for the fall of the Republic and emergence of the Caesars that followed him, curtailing many of the powers and privileges of the Senate. By his exploits did Rome become feared in all the courts of the Eurasian-African nations, no kingdom inaccessible to Italy. As he was infamous, so too 552 years after his birth did an infamous invader attack Italy in 452 AD named Attila the Hun, from regions so far away that the Romans had never heard of them. But this was not the first shock Rome had felt in the early Anno Domini years.

On August 24th, 410 AD at midnight the Salarian Gate of Rome was quietly opened and Alaric the Goth led the Visigothic army into the Eternal City, initiating the infamous Sack of Rome. These Germanic hosts invaded the city and ransacked its private and public buildings, treasuries, galleries, storehouses and temples. This emotional and humbling event inspired Augustine to compose his renowned work named *City of God*, in which he noted with awe that never before had an invasion force ransacked an entire city, leaving all of its Christian shrines and sanctuaries *untouched*. Why this amazed Augustine is truly the mystery, for Christianity had early on swept through the German kingdoms at the borders of the Roman provincial territories like an uncontrollable fire. In fact, the spirit of Christian Germania would go on to forge the history of Europe itself. This Sack of Rome was 552 years before the founding of the Holy Roman Empire of *Germany* which began when Otto I was crowned Emperor by Pope John XII at Nuremburg. We should not be surprised to learn that the Holy Roman Empire adopted as its seal the Double-headed Phoenix. (7) This alliance between the Papacy, Italy and Germany would reoccur in World War II.

In the year 762 AD the Abbasid Dynasty makes Baghdad its capitol, the seat of Muslim authority, effectively establishing a *second* Babylon. The year 1314 AD was 552 years later when the whole of the Middle East, Europe and Asia reported unusual astronomical signs and disturbances. European historians reported a "...great black darkness," in the night skies (8) while on the other side of the planet during the daytime the Chinese recorded that the *sun darkened*. (9) This date was one of meteoric rain, plague fogs, strange mists and terrible earthquakes. This was not planet Phoenix. The event and phenomenon is fully expounded in *Anunnaki Homeworld*, for this was the last time the Anunnaki planet NIBIRU passed through the inner system before it returns during the coming "Apocalypse." One of the most distinct denominators between these events is the fact that Babylon is the biblical symbol for an *oppressive religion*, and this symbol often refers to both Islam as well as the Papacy of Rome, which assumes itself to be the true authority of the Christian world. Which, of course, is not true. It was in 1314 AD that Jacques de Molay was burned at the stake, officially ending the Knights Templar, an organization that had protected Church interests for many years but fell to the greed and envy of a conspiracy between Papacy officials and King Philip IV of France. Philip had ruthlessly tortured to death 36 Knights, burned 54 of them at the stake and confiscated all their estates and other wealth he could find. This oppressive nature of the Church leads us to our next 552 year period.

As 1314 AD was 552 years after 762 AD, so too does 1314 AD begin a 552 years timeline to 1866 AD, when the major European kingdoms of *France*, Spain, Austria and others renounced Roman Catholic Papal authority and the sovereign Pope within the Vatican was *stripped of his power* over most of Europe. It had been earlier prophesied that the Papacy would fall in this year of 1866 AD, centuries before the event by the chronologist Heidelburg, as well as Cellerius, Pareus, Gresner and Fleming – all having specified the year 1866 AD. (10) Also of interest concerning this unique year is the fact that it was exactly 3312 years (552 x 6) after the Exodus of Israel from Egypt in 1447 BC after the Ten Plagues devastated Africa. These 3312 years parallel the entire longevity of the Age of the Phoenix. In 1447 BC the volcanic mountain of Santorini exploded and caused tsunamis, earthquakes and severe atmospheric contamination, and in 1866 AD there was *again* volcanic activity at Santorini in the Aegean.

One of the most unusual 552 year alignments again involves Islam and the Hebrews. In 1356 AD a major earthquake shocked Egypt and destroyed the final vestiges of the Pharos Lighthouse, an immense structure wherein was composed by Hebrew scholars the famous Greek Septuagint translation of the *Torah* (Five Books of Moses) in the 3rd century BC. The quake dislodged some of the gigantic white polished facing blocks on the original exterior of the Great Pyramid, exposing for the first time the core inner limestone blocks of the monument that ascend 203 levels upward to its empty platform 481 feet above the plateau. These massive 20-ton casing blocks were smashed off the pyramid as local Muslims mined the Giza site for building materials to rebuild the earthquake-shattered city of Cairo. (11) This earthquake was 552 years before another series of disasters and quakes that transpired in 1908 AD.

In 1908 AD over 900 witnesses watched a bright light trailing flames traverse the skies over Siberia. The object exploded in the air and destroyed 300 square miles of dense woods over Tunguska, Siberia, even knocking people off their feet 100 miles away. The pressure of the blast affected barometers in England and sunsets for weeks were made beautiful by atmospheric dust. Radioactive traces of iridium (mostly an extraterrestrial element) from this comet reveal that the object weighed about seven million tons. (12) This event occurred on June 30th.

On December 28th at 5:20 AM in 1908 AD a major earthquake rocked Messina, Sicily, followed by a tsunami 26 feet high that wreaked destruction through the city and pulled many people out to sea as the waters receded. Gas lines ruptured and fires raged uncontrolled for a total death toll of 120,000 people. (13) These earthquakes were 552 years apart. The link between the Israelites and Egypt is made more profound by a startling archeological confirmation made in 1908 AD. After an archaeologist named Naville in 1883 AD claimed to have found evidence proving the Exodus story in the walls of the Egyptian ruins of the city of Pithom, another archaeologist named Kyle traveled to Egypt and researched the site himself in 1908 AD. He concluded that the evidence was real, that the walls of the city of Pithom exhibited at the lowest courses of masonry normal clay brick straw content, however, the middle courses were deficient in their amounts of traces of straw and the uppermost courses had *no straw at all*. (14) This was confirmation of the biblical Exodus story where we learn that Pharaoh in Egypt made the Israelites erect structures with clay-burned bricks, shortening their straw rations over and over again.

A further parallel of this scenario is found when considering that the Muslim extraction engineers were removing the *bricks* (stone casing blocks) off of the Great Pyramid to rebuild their buildings and mosques, the Great Pyramid site known biblically as *Goshen*, where the *Israelites in Egypt lived*. They could see the enormous pyramid every day and it was common knowledge of the times that the structure had been erected by their own *Sethite* ancestors before the Flood. The story of the origin and purpose of the Great Pyramid is covered extensively in *Lost Scriptures of Giza*.

Not surprisingly, our final 552 year period is also the *last* 552 year epoch possible in the 6000 year timeline of man's banishment. It was the year 1554 AD when a *comet* recorded by European astronomers passed through the night skies. This comet appeared *552 years* prior to the year 6000 Annus Mundi (2106 AD). 1554 AD is the year 5448 AM, this being the precise *height of the Great Pyramid* in Pyramid Inches: 5448. This 552 years measures the time/length until the *Chief Cornerstone* will descend from heaven and *remove the curse* that had afflicted mankind since the year 3895 BC. He will for His millennial kingdom reunite all the descendants of Israel and Judah as well as Abraham and all those grafted into the Divine Family by faith. And again, we are not surprised to find that 1554 AD is exactly 2484 years (414 x 6) after the split between Israel and Judah in 931 BC at King Solomon's death. This 552 years/inch dimension is clearly found as the width of the ceiling of the Subterranean Chamber deep below the Great Pyramid as seen in *Chronotecture*.

Our review of some of the historic Phoenix Cycles complete, we now delve into the archives of history as it unfolded through the patterns of *Cursed Earth* periods.

V

Cursed Earth Periods

*Have you then deciphered the beginning,
that you ask about the end? For where the
beginning is, there shall be an end ... blessed
is the man who reaches the beginning, he
will know the end.*

—Words of Jesus, *Gospel
of Thomas*

A Cursed Earth period is 414 years, a timeframe easily detected through the narratives of the Old Testament. One such example begins in 2235 BC (1660 AM), four years after the Flood. Noah was drunk from the produce of his vineyard, both he and his wife Naamah lying naked. His youngest son Ham (Chem) sees this and, captivated by the beauty of his own mother, lies with her. His brothers Japheth and Shem are made aware of the trespass, and in Genesis it reads that they walked backwards with a blanket to cover her so as not to view her body. They took this precaution because under patriarchal law it was a *curse* to see your father's nakedness, which Levitical law explained simply means looking upon the exposed body of your father's wife. When two sexually unite in marriage they become *one flesh*. The trespass was against Noah and he discovered the matter when Naamah is found pregnant. She gives birth to *Canaan*, child of the Curse of Noah –

> Cursed be Canaan; a servant of servants shall
> he be unto his brethren. And he said, Blessed
> be the Lord God of Shem; and Canaan
> shall be his servant. God shall enlarge
> Japheth, and he shall dwell in the tents of
> Shem; and Canaan shall be his servant. (1)

Canaan's progeny bore this curse. He became a patriarch over seven tribes that grew into the populous Canaanite nations, who against the advice of Ham, violated the Law of the Lots and took residence in a land belonging to

41

Shem and his descendants. This land was in turn called Canaan, the allotted inheritance of the Israelites. In the year 1821 BC the Philistines stole a well belonging to Abraham, this being the *414th year* of the curse of Noah. The Philistines were descendants of Ham. As retainers Abraham had among his servants the huge Anakim giants famous in Scriptures as well as many friends and allies and, as Genesis 14-15 clearly reveals, he was capable of making war against Philistia. But their king, named Abimelech (royal title of Philistine kings) profusely apologized for the trespass of his shepherds, offering gifts and a peaceful settlement. This story is recorded in Genesis and the Book of Jasher. (2) To remove a boundary stone or to steal a well was punishable by death and adhered to as a sort of international code of statutory ethics.

What Abimelech did not know was that his people were under a curse and that they had even established the means by which they were to suffer under it, however, this means was *averted* by the wise actions of King Abimelech of Philistia. The curse was not dispelled, but *delayed*. The Philistines were saved by wisdom, however, Abraham in that year was also afflicted by the Canaanites who also illegally occupied the land. And unfortunately for them, the curse was made firm against the whole of the cities of Canaan.

Exactly 414 years after Abimelech's wisdom rescued his people from a judgment of God, Abraham's descendants marched across the Jordan River at the death of Moses, but led by Joshua, and first destroyed the Canaanite city of Jericho, then Ai and then over a hundred others in 1407 BC (2488 AM). The Canaanites with their Amorite allies gathered to repel the Israelites in a massive host that was immediately routed and killed off by meteoric rain, earthquake, the unleashing of *evil angels* upon these Hamitic peoples and the continual slaughter from the blades of Israel. As Joshua directed the course of the battle for Canaan the sun stood still in the sky, and the moon ceased moving for a period of about 20 hours, as the Israelites routed and hunted down their Canaanite enemies. This was two Cursed Earth periods (828 years) after the Cursed of Noah. The solaric standstill will be detailed later in this book, as it related specifically to planet Phoenix. Interestingly, as their Canaanite neighbors were being eradicated, the Philistine cities along the coast were left *unmolested* by Israel, enjoying a temporal immunity from the curse.

This conquering by the Israelites was the Conquest of Canaan 40 years after their Exodus from Egypt in 1447 BC. One Cursed Earth period of 414 years after this was the year the very first king of Israel, Saul, became cursed by God in 1033 BC (2862 AM). This is how this story unfolds: In 1447 BC the

Israelites, led by Moses, were moving away from a cataclysmic-torn Egypt. They were passing around the domain of the Amalekites after Moses was specifically instructed by God to stay away from the region of the Philistines. Evidently they were still receiving immunity from the curse. As the train of over 1.5 million people traversed these wastes, Agag of the Amalekites had his soldiers attack the defenseless and scattered caravan, taking loot and prisoners he never returned. In a rage, Moses *cursed* the Amalekites. (3)

Precisely 414 years later the Israelites were at war with the *Amalekites* in 1033 BC, who were of the progeny of Ham, and through the prophet Samuel, God instructed King Saul to kill them all, and to execute Agag (regal title of Amalek). But when Saul captured Agag he is taken aback by the enormity of the man, a human colossus, for Agag was a descendant of the giants. He must have been truly heroic in stature because the Scriptures clearly read that Saul himself was a head taller than the tallest among Israel. Despite knowing that Amalek was cursed by their venerable ancestor Moses, Saul refused to slay Agag, fully taking upon himself the *curse*. Samuel the prophet, to avert a national disaster (transference of the curse upon Israel), took up the sword and executed Agag, making the curse upon King Saul *personally*. (4)

The king of Israel was now bearer of the curse; his reign was instantly plagued with a series of defeats in a constantly returning war against the *Philistines*. For 22 years the Philistines oppressed Israel, led by a military aristocracy of Giants and their offspring who had escaped Canaan in 1407 BC when the Israelites were routing the Canaanites. The Scriptures read that the Rephaim and Anakim Giants took up residence among the Philistines in 1407 BC. In the days of Saul the principle threat was the giant Goliath of Gath who had previously even stolen the Ark of the Covenant. But later the Philistines voluntarily returned the Ark to Israel after it was blamed for an extensive epidemic. After the passage of 22 years of unceasing conflicts, Saul and his son Jonathon were killed by the Philistines in 1011 BC after Saul reconfirmed his curse by not consulting God in his trouble, but the Witch of Endor. (5) It was none other than the spirit of the prophet Samuel that visited him under the divination of the witch, informing Saul of his curse.

Saul was Israel's first king, and like their leader, the people too rebelled against the ordinances of God. Among the cities of Israel were surviving remnants of the Canaanites, who were supposed to have been killed in consequence to the curse and their violation of the Law of the Lots. To learn more about these patriarchal and ancestral covenants, read *King of the Giants: Mighty Hunter of World Mythology*, to be released by this author. The people lost their first king, 1011 BC, and 414 years later in 597 BC the people of Israel lost all their original sacred Temple artifacts and relics when the Babylonians sacked Jerusalem and the Temple precinct. This resulted in the loss of 1000 golden cups, 29 silver censors, 1000 silver cups, 30 golden vials and 2410 silver vials and 1000 other vessels. (6) These same holy vessels are what *cursed* Babylon in 537 BC when King Nabonidus, son of Nebuchadnezzar II, used the holy Temple vessels in a palace banquet party (some say orgy). The Finger of God appeared and wrote MENE MENE TEKEL UPHARSIN on the wall and the exact same night after Daniel interpreted this to mean that Babylon would fall to the Medes and Persians, the city of Babylon was taken by surprise without a fight and King Nabonidus was killed.

The Temple relics are related to the Cursed Earth system in another profound way. Again, they are associated with Babylon. And in a most startling way our history here interweaves through the tapestry of time back to Nebuchadnezzar II and his son Nabonidus. The Temple relics, artifacts and the foundation of the Temple itself were made and laid in 967 BC (2928 AM) by King Solomon, this being his 4th regnal year over Israel. (7) This is exactly 480 years after the Exodus in 1447 BC as described in Scriptural chronology. (8) King Hiram of Phoenicia provided lumber and metalworkers for the holy project. This 967 BC year began a 414 year timeline to the *healing of Nebuchadnezzar II* and the first regnal year of his son, Nabonidus, in 553 BC. Nebuchadnezzar had previously, seven years earlier (the Temple required seven years to build) insulted the Creator and was specifically *cursed* by God to suffer lycanthrope, to become mentally deranged to the extent that he assumed canine and bestial characteristics, forgetting his own identity for a period of seven years. In the 414th year the king was healed of his condition. The Babylonian court officials and nobility kept the matter private and confined Nebuchadnezzar II for the designated period, having been assured by the prophet Daniel, high in the king's court, that the condition would pass. King Nebuchadnezzar's remarkable testimony of the God of Daniel is preserved in the Book of Daniel.

Many other examples of the Cursed Earth system abound. In the year 2291 BC the seventh Anunnaki king assumed the throne at Shurrupak before the Great Deluge, named Ubarutu. He oppressed the people, introduced unprecedented evils and his reign serves as a prophetic model of the draconic rulership of the Antichrist during the Apocalypse. The Anunnaki are *cursed* and have made a covenant with the Evil One. Exactly 414 years after the emergence of Ubarutu (whose reign was ended by the Flood in 2239 BC),

Abraham, patriarch of many peoples, entered into a *covenant* with God in 1877 BC. God promised Abraham that he would be the father of many nations and that his progeny would be as numerous as the stars of heaven. He was also informed that the beginning of the fulfillment of this promise would result in 430 years that his people would be in bondage under Egyptian authority. (9) God commanded the patriarch to travel to Egypt in this same year, thus beginning the 430 years to the deliverance of his descendants (Israel) in the Exodus led by Moses in 1447 BC.

The global diluvian disaster that ended the reign of the seventh Anunnaki king in 2239 BC began itself a 414 year period to 1825 BC. This was the year that Abraham had concluded his 144th month (12 years) in Egypt translating the writings upon the four faces of the Great Pyramid, which was his second trip to Egypt. Abraham was taught how to interpret the ancient Sumerian texts from before the Flood, taught by none other than the ancient *Noah* before he died. This subject of the pyramid's writings is covered in *The Lost Scriptures of Giza*. This was also the year that planet Phoenix passed through the inner system, being the year 2070 Annus Mundi (414 x 5). In this year Abraham was providing hundreds of scholars from around the world, detailed historical, prophetical and wisdom literature translated from the pyramid's writings, knowledge that would later become the core foundational data, spread throughout the world's oldest scientific and religious books. This college of dissemination was ordained by God and is the reason why so many civilizations in antiquity having no contact with one another, maintained the same concepts, histories, and beliefs about the future.

In 1899 BC the empire of Mesopotamia seated at Akkad and ruled by Sargon I was shattered in a devastating earthquake and lightning storm from heaven as the Dark Satellite (see *Anunnaki Homeworld* for orbital history) nearly collides into earth. The Akkadian ruler was almost finished with a gigantic pyramidal structure known to us as the Tower of Babel, built by the aid of Anakim giants known in traditional lore as the Titans. The people and their cultures fragmented and many whole groups migrated to distant parts of the Earth, fleeing the tyranny of Sargon I as well as seeking newer pastures to build their own cities. It was 414 years after this episode that the Ephraimites, descendants of Joseph, patriarch of the Israeli Tribe of Joseph, son of Jacob, prematurely attacked Canaan even after they were warned that it was not the *allotted time*. These overzealous warriors were virtually eradicated, 30,000 men of Ephraim with only 10 survivors, by none other than the *Philistine* army. (10) Also in this year Caleb was born, who would later be one of the 12 spies of Israel – and one of the only two who brought back a faithful report.

In 1639 BC the patriarch of Israel, Jacob, died. Before his death in Egypt, on his deathbed, he summoned Joseph and his two sons Manesseh and Ephraim. Joseph at this time was Grand Vizier of all Egypt under the Hyksos Dynasty seated at Memphis. Inspired by God in conformance with the Abrahamic Covenant, Jacob adopted Joseph's sons as his own. These two tribes effectively became one, the 13th Tribe of Israel, Tribe of Adoption. This year was 414 years before the sun darkening of 1135 BC when the Israelite descendants of the Tribe of Dan (which is later excommunicated from the family of Israel) engage the Fomorii giants in ancient Ireland in the Battle of Magh-Tureadh, the Fomorii being kin to the *Philistines*. It was also during the reign of Nebuchadnessar I.

The amazing synchronicity of parallels and cycles continue here in a remarkable way. The 1135 BC sun darkening by the transit of Phoenix began itself a 414 year countdown to 721 BC when the Assyrian King *Sargon II* deported the remaining tribes of Israel, among them the descendants of Manesseh and Ephraim, who would later migrate deeper into Northern Asia and Europe, Asia Minor and eventually cross the English Channel into the Pretainic Isles (ancient Albion: Britain) and the Atlantic Ocean, founding the 13 Colonies that would emerge into the United States of America. This deportation began the fulfillment of the Abrahamic Covenant – that his seed would become many mighty nations.

The deportation of Israel (while the kingdom of Judah remained intact) began in 745 BC with Assyrian king Tiglath-Pileser, known biblically as Pul. (11) This deportation was amazingly 2070 years (414 x 5) *Anno Pyramid*, or, 2070 years after the completion of the Great Pyramid in Egypt at Giza in 2815 BC (1080 AM), which is fully covered in *Chronicon: Timelines of the Ancient Future*. The Deportation begins the *Post-Exilic Chronology*, a countdown of 2520 years (360 x 7) to the founding of the United States of America, the Last Days *Empire of Adoption* in the western land of promise, in 1776 AD. Amazingly, the 414th year of the Post-Exilic Chronology was 331 BC, the exact year that the Anunnaki planet NIBIRU entered the inner system from the Deep (nether region of solar system below the ecliptic) at the exact same time Alexander the Great's army from Grecia defeated the Persian forces in the *Battle of Gaugamela*, routing King Darius and beginning western imperialism.

As previously related in our review of Phoenix Cycles we learned that in 1687 BC the sun darkened while the sons of Jacob battled the Canaanites. This event was caused by the transit of Phoenix. Counting 414 years later we arrive at 1273 BC, the start of the Assyrian Empire with the annexation of Babylon. Phoenix passed through the inner system in this year and the Assyrians adopted the Winged Disk as their Great Seal. Counting yet *another* 414 years we come to 859 BC. Evidence taken from Assyrian seals and palace reliefs show that the Assyrians were *expecting* the return of the Winged Disk (Phoenix), according to Zecharia Sitchin. (12) The Phoenix was the antediluvian sign of *kingship* from the days of Enoch. In this year King Shalmaneser III succeeded his father Ashurnasirpal II, who had previously captured the coastal cities of Old Tyre, Sidon and Byblos (Gebal). The famous Black Obelisk of Shalmaneser III depicts a relief of the Israelite king Jehu, also named in the Scriptures, kneeling before the Assyrian monarch and paying tribute. Above Jehu is the Winged Disk of Phoenix.

This year of 859 BC was during the reign of the 32nd king of Cuzco in Peru. Montesinos, the historian of the Andes, wrote that the 32nd king of Cuzco reigned exactly 2070 years (414 x 5) after the start of Peruvian Reckoning. (13) This exact number, 2070, denotes a precise knowledge of the Phoenix system (15 orbits of this planet). This date is 548 years after the 1407 BC Conquest of Canaan (15th king of Cuzco), with a period of 17 kings reigning on average for 32 years each. This 2070 year Peruvian timeline began in 2929 BC, exactly 20 years before the end of the reign of Enoch, who is remembered as the most famous of all *chronologists*. Of course, in 2929 BC planet Phoenix passed through the inner system.

With Assyria and Media so prevalent in our studies of Phoenix Cycles and Cursed Earth periods we are not astounded to uncover more alignments involving both. In 615 BC King Artaxerxes of Media began a war against the Assyrian Empire and 414 years after this war began, the Punic War between Rome and Carthage ended in 201 BC. The Carthaginians were the descendants of the Phoenicians, and both Rome and Carthage were the Mediterranean superpowers in 201 BC just as, in the Middle East in 615 BC, were Media and Assyria.

Just one year after the start of the Medo-Assyrian War of 615 BC, in *614 BC,* Median forces destroyed the Assyrian city of Ashur and the Medes forged an alliance with Babylon. And 414 years later Rome declared war and marched against King Philip IV of Macedon in their struggle to dismantle the Macedonian kingdoms. Two years into this Medo-Babylonian alliance against Assyria, the Assyrian cities of Nimrud and Nineveh were laid waste under *Nebuchadnezzar II*. This too began a 414 year countdown to 198 BC, when Roman forces laid waste to Macedonian citadels and cities in their war against King Philip IV.

As we begin to conclude our review of the Old World through the window of Cursed Earth time, it is important that we look back to the beginning of the Cursed Earth system which began in 3895 BC, with man's banishment from Paradise. The 2070th year of the Annus Mundi timeline ended 414 years after the Great Flood in 2239 BC, or 1825 BC, which was the year Abraham was translating the Giza Texts from off the surfaces of the Great Pyramid. While this may appear innocuous, there is a much deeper relation between these two events, separated by the span of 2070 years (414 x 5). This startling connection concerns the *Word of God*. In 3895 BC (Year One AM) mankind was ousted from Eden because he allowed a *translator* (serpent) to explain to him the Word of God (Tree of Knowledge of Good and Evil), rather than accepting it obediently. Humanity then discerned that there was a difference between good and evil and this revelation caused people to doubt themselves and condemn everything they had formerly enjoyed, actions previously *sanctioned* by God because mankind was faultless. Mankind was *cursed*. Cut off from the Word, men wandered the earth, multiplied, built their cities and civilizations and refused to listen to those prophets and priests (the *interpreters* man chose) that warned them of the coming global destruction of humanity and earth itself (the Flood). Now, 414 years after this cataclysm, humanity had rebounded and was multiplying rapidly. News spread abroad of the discovered monument from before the Flood that exhibited the original writings of the Word of God, found half buried in the sands of Egypt. This was the Great Pyramid of Giza, which had remained beneath the waters of the Mediterranean along with the entire Delta region of Egypt's coast for 340 years after the Flood until 1899 BC. A massive quake shoved this area upward, which is why Egypt was referred to at this time as The Raised Land. Mounds of sand and seashells were removed and sages and scholars from as far as India traveled to Egypt to hear the prophet and scribe Abraham (later to be remembered as Brahma) read his *translations* to the Word of God found on the casing blocks of the gigantic pyramid. These translations became the core source materials for the oldest religious and mythical texts in the world. It was in 1825 BC, Abraham's 12th year translating these immense writings, that men yielded to the Word of God, exactly 2070 years after man first succumbed to the words of an evil being that deceived him. *Lost Scriptures of Giza* serves to cover all of this history.

The period of 2070 years (414 x 5) is 207 x 10 years, not merely a mathematical fact but an astronomical one. As 207 years is half of a 414 year Cursed Earth period, we are confronted with an intriguing fact: in 2446 BC (1449 AM) a five-planet alignment of Mercury, Venus, Mars, Jupiter and Saturn formed a ladder in the sky as recorded by Chinese astronomers. (14) This planetary alignment was *207 years* before the Deluge in 2239 BC (1656 AM). The Deluge ended an epoch, the Antediluvian World, and the year was the synchronization between the Phoenix Cycle and Cursed Earth systems, for this year was 552 x 3 and 414 x 4. But 207 years here, at this time, infers that the pattern is incomplete. And this is indeed the case. Though this is not the subject matter of this work, this author has written extensively on the Pre-Adamic World and its destruction in the year 4309 BC in his *Descent of the Seven Kings* and huge tome entitled *Chronicon*. The ruination of the Pre-Adamic World actually begins the Cursed Earth Chronology, and as we will find toward the end of this work, is a calendar intimately connected to the Great Pyramid's *Giza Course Countdown* calendar that sequentially counts down Earth's final 204 years until the return of the Chief Cornerstone in the year 6000 (2106 AM). This Pre-Adamic catastrophe happened precisely *414 years* before mankind was banished from Paradise, which began the 6000 years of his exile. Thus, from the end of Pre-Adamic Earth to the end of humanity at the Great Flood, there passed exactly *2070 years* (414 x 5).

Within the context of these historical timelines is unveiled the secret to the orbital length of planet Phoenix. The great 552 year Phoenix Cycles began with the Curse of man in 3895 BC when he was ousted from Eden, but the 414 year Cursed Earth timeline is more ancient, not starting in 3895 BC (Year One AM), but *414 years* before, in 4309 BC, with the end of Pre-Adamic Earth.

It is imperative that the reader recall this fact later in this study – that the Phoenix Cycle system is specifically *prefixed* with a 414 year Cursed Earth period. . . for the end always lies in the beginning. Our review of the Old World finished, let us now peer into more contemporary times.

VI

Modern Cursed Earth Periods

*The gods give us signs of future events. If we
go wrong about them it is not the divinity but
men's interpretation that is at fault.*

—Cicero, quoted by Ammianus
Marcellinus (Book 21)

In the first century Anno Domini (AD) the Roman world was the apex
of nations. Its engineers and architects were unrivalled, artists were brilliant
copyists of older Grecian masterworks, its statesmen were compelling orators
and historians, its warriors being the legionnaires, the knights and Praetorians
with their traditional Trojan-styled apparel. The whole wide range of human
endeavor was at its height and Rome was the epicenter of this activity. What
we boast today as contemporary invention and innovation was already
precedent at Rome. Even the bikini bathing suit, so fashionable in the present,
is but a shadow of the beachwear of Roman times as seen clearly upon the
frescoes excavated from Pompeii from underneath the detritus of Mount
Vesuvius. This terrible volcano that buried the settlements of a pre-Etruscan
race provided Rome with a sturdy foundation for their own unwitting cities,
Herculaneum and Pompeii, resort luxury-cities wherein were enjoyed all the
vice and vanity the affluent could afford. Brothels and sex slaves, gambling
and taverns, all the best money could buy was enjoyed when the mountain
again buried a civilization in 79 AD, and the result was also the same as in
1687 BC. . . the entire region was buried and newer topsoil and flora provided
a new landscape for fauna and people who built towns and villas over this
area.

But Rome had been warned, though no one was watching the signs. The
Cursed Earth system of 414 years also applies to *days* in the longevity of
individuals and cities. Mount Vesuvius did not end the lives of so many, and
so violently, without first providing the inhabitants of the land a warning.

This warning came on February 5th, 62 AD, when Vesuvius trembled and
Herculaneum and Pompeii suffered from an earthquake. (1) Then, after the
passage of 6210 days, in 79 AD, Vesuvius unleashed the wrath appointed
and the Roman cities suffered the effect of their curse with the burying alive
of Pompeii and Herculaneum, as well as the sorrowful demise of our hero-

51

historian Pliny the Elder. The warning? It is found in the lapse of 6210 days, for this is *414 x 15* days, which parallels the entire length of the Cursed Earth Chronology in *years*, as we will learn in this work. But this is a far cry from the really sobering reality of the 62 AD earthquake in the Roman dominion.

The 62 AD quake began a Cursed Earth period against all of Rome. This was the exact year, 414 years later, of the Fall of Rome in 476 AD when Odoacer the German defeated the boy-Emperor Romulus Augustulus, a most famous accomplishment that altered the demographics of the entire Roman-occupied world and Europe. Rome's end mirrored its beginning, for as a multitude of early Roman historians attest, the Romans were descendants of the Trojans that lost out in the Trojan War against the Mycenaeans and their Greek allies. The Trojan War was in 1229 BC (see *Chronicon*), or *476 years* before the city of Rome was founded by a young king named *Romulus*, paralleling the Fall of Rome in *476 AD*. The synchronicity is so profound that that the majority of scholars attribute Rome's origin to a mythical past, despite all the ancient evidence.

The descendants of Troy integrated socially through mass marriages with the local Etruscans of Semitic stock, as well as the Latins also in Italy. Through Rome's history an influx of people descended from the Israelite tribes deported by Assyria in 745 BC and again in 721 BC joined the Roman peoples, as well as many from Judah. During the centuries of imperial dominance that Rome enjoyed, their official seal and standard was of the Phoenix, and only in later Roman times was it changed to the Eagle. When Rome fell, the descendants of the Israelite tribes that had merged with the Cimmerians, Scythians, Goths and the Celts (Gauls) had become independent cultures and populous throughout Europe and Asia Minor. The Fall of Rome *began* the emergence of the *Ten Kingdoms* which were prophesied to come by Daniel, people described in Scripture as *iron mixed with miry clay* (Japhetic-Semitic blood). Rome itself was the Dreadful Beast, also called the Iron Empire.

The transition from Phoenix to Eagle denotes an empire in decline. But this decline not always refers to a nation's might, but infers more of a *moral* decline. The values of early Rome when it was a Republic were no longer upheld and the imperial attitudes toward other cultures had degenerated into a hardline master-slave relationship. Which leads us to another strange coincidence in the history of planet Phoenix. Roman civilization was supported largely by *slavery*, for slaves were owned in the majority of households and by virtually all land owners. Often the transaction of deeds for land also meant that all slaves attached to the regions were purchased or sold. Slaves were used in commerce and industry and even trained in academics. Even the Legions had slave trains – people in bondage who served the soldiers by cleaning, cooking, providing weapon maintenance and sex. As planet Phoenix traversed the inner system in 246 AD, as it continued its ancient 138 year orbit, the Goths flooded the Roman provincial region of Dacia in a full-scale invasion, and the Dacians were no more heard from in history. Be it either assimilation with the Goths or genocide, the Dacians loyal to Rome vanished. (2) The incursions of the Germans into Roman-occupied lands struck an intense fear into all landholders and Romans, but were met with cheering and jubilance by the immense slave populations that regarded the Goths as *liberators*.

The Ten Kingdoms of Europe were a political extension of the authority of Rome, however, they were sovereign people actually listed by Niccolo Machievelli (who never cited nor hinted he was aware of the prophecies of Daniel). Machievelli wrote that the Ten Kingdoms emerged with the Fall of Rome: the Lombards, Burgundians, Franks, Ostragoths, Visigoths, Vandals, Heruli, Suevi, Huns and the Saxons. (3) These kingdoms would contend with each other for supremacy for 1300 years. These offshoots of the Ten Tribes of Israel who had been forcefully integrated by the Assyrians with the Indo-Germanic tribes of the Scythians, Cimmerians and Gimry (who were also under the yoke of Assyria) were to provide the Last Days *Iron Empire* with its stock of peoples, a mixed nation that would fulfill the role of the Empire of Adoption (13th Tribe of Israel adopted by Jacob). This 1300 year period, from the Fall of Rome by German invasion until the founding of the United States of America out of 13 colonies in 1776 AD, fulfills the prophetic timeline of the emergence of the Iron Empire – the United States founded by Anglo-Saxons, descended from the *German* tribes of the Angles and Saxons (sons of Isaac). And the original Seal of the United States was none other than that of *Phoenix*. Remember this fact, for it will prove even more astonishing later in this book.

The center of learning in history, astronomy, geography, philosophy and foreign theologies was at Alexandria, Egypt, famous city of the Macedonian kings. In the year 389 AD, Christians persecuted the scholars of the famous Library there, a ploy designed by the religious propaganda of a local bishop with a personal vendetta. Men under pretended religious indignation at the assumed offenses of a famous female mathematician and philosopher named Hypatia, dragged the woman off of her chariot and violently stripped her naked before the people, parading her through the streets like a criminal until men gathered in a mass with oyster shells and, while holding her down, others flayed her alive inside a Church. Many priceless books were burned and lost that had not yet been hidden by the stewards. It was events like this that led to the Dark Ages, when European nations were plunged into the abyss of ignorance. Counting 414 years into the future after this persecution of knowledge and the knowledgeable, the seat of learning had fallen from the west and had, almost overnight, emerged back in the east, in Baghdad in Babylonia, among the Islamic scholars who inherited many of the books from Alexandria that Muslim invaders had taken – men hungering for the learning that Europeans neglected. The treasures of Alexandria at this time were dispersed by Muslim merchants along the trade routes and found their way into the courts of many European kings. In 803 AD the Islamic empire seated at Baghdad had among them the most influential scholars in the world.

Further signifying the darkening of western intellectualism and the rise of Islamic light was Alexandria's reconfirming of its curse in 415 AD. In this fateful year the Archbishop of Alexandria had the Library raided by his Christian agents and most of the books not approved by the bishop that were not yet secreted away were burned. Reconfirming the curse that fell upon Alexandria in 389 AD, the bishop ordered that all of the scholars share the same fate as Hypatia. In accordance with his mandate, the historians, scholars, mathematicians, geographers and stewards of the Library were stripped and flayed alive using oyster shells. (4) This second event was 414 years before 829 AD when the Caliph Al Mamoun had built Islam's first astronomical observatory, and at *Baghdad*. Al Mamoun was a scholar himself who appreciated learning and had the first Islamic chart on astronomy drafted. Additionally, it was this same man who in 820 AD tunneled into the Great Pyramid at Giza and discovered the formerly unknown and hidden *Ascendant* Passages and Chambers. This year, 829 AD was *207 years* (half of 414) after the establishment of the Islamic Hijrah Calendar in 622 AD. As this is half a Cursed Earth period, we are not surprised to find that 829 AD begins a Cursed Earth period to 1243 AD, which began the final year of the Crusades of the Christian-Muslim Wars over Jerusalem, which climaxed in 1244 AD when the Islamic armies took Jerusalem for the final time. After this, both the Muslim and Christian worlds would suffer a curse. . . the inclusion of the rampaging Mongols that would crush the Abbasid Dynasty at Baghdad and make excursions into Europe leaving blood and ashes in their wake.

The oppressive influence of the official Church over the European kings made the Crusades possible. The first Roman Imperial *Christian* Emperor was Constantine the Great, who had put an end to the religious persecutions of Christians and moved the seat of Roman power from Rome to the older Greek city of Byzantium. It was this figure in history who gave recognition to the Christian church, which is why 414 years later, in the year 751 AD, a Cursed Earth period after Constantine's death in 337 AD, the Roman Church enforced the *Donation of Constantine*. This was a document produced by the Church allegedly written by the hand of Constantine before he died in the year 337 AD. This document claimed that on his deathbed Constantine gave the entire power and wealth of the Roman Empire to the Church, including the right to crown and dethrone kings. Though many believed the document to be a fake, the Church had become so powerful that this move went virtually unopposed. This Church takeover of the courts of Europe led to the first move against the royalty of Europe, an attack against the Merovingian Kings of France. The Merovingians, deposed by treachery, were particularly despised by the Church officials because these French kings had previously disallowed the Church to interfere with all affairs of State. The Merovingians claimed direct ancestry from *Israel*, particularly from the descent of the kings in line with Solomon and David. (5) In 751 AD the Divine Right of Kings became the Divine Right of the Church, and the entire course of European history would have been drastically different had not the European monarchies allowed the papacy to control state affairs. It has now been proven by Lorenzo Valla that the document, Donation of Constantine, was a fraud from its inception for it was written using biblical passages quoted from the Latin Vulgate version of the Bible. The Vulgate had been compiled by St. Jerome, who was born *two decades after* Constantine's death. The document had to have been composed about 50 years after Constantine died. (6)

After the Mongols came, it was the turn of the Turks to terrorize Europe. In 1062 AD the Turkish armies invaded Greece and virtually destroyed Greek Christendom as well as northwestern Italy. This began the Turk offensive against Europe. During this Cursed Earth period of 414 years, beginning at this date, the world endured changes and disasters on a global scale tantamount to those we reviewed concerning the 552 year Phoenix Cycles: 2239 BC, 1687 BC, 1135 BC and 583 BC. In 1066 AD, according to Geoffrey Gaimar's *Lestoire des Englis*, the people from England witnessed a fire in the sky that burned brilliantly, which came near to the Earth, appeared to move erratically, then descended into the sea. In many places the forests caught fire. (7) Following this comet the Normans appeared on the shores of England and began, as the overlords of the Isles, forever changing the English culture. In 1084 AD the Normans even sacked Rome. In 1095 AD Pope Urban II called for a Crusade against Islam and, in 1096 AD, Peter the Hermit answered the call by leading the First Crusade. In 1099 AD the Crusaders of Europe butchered those they came upon, instead of delivering the inhabitants of Christian Antioch, Edessa, Jerusalem and Trilopi in Libya. In 1138 AD a quake killed 230,000 people in Aleppo, Syria and comet activity increased alarmingly in 1150 AD. In 1169 AD Mount Etna in Sicily erupted and an asteroid impacted the moon in 1178 AD. (8)

From 1194-1202 AD flooding, quakes and plagues decimated Asia, the Middle East and Africa. In 1209 AD Pope Innocent III initiated the horrific Inquisition that would result in the hideously painful deaths of over a million truly innocent victims of Church tyranny. In 1214 AD the Mongols invaded China, while in England the following year King John enacted the Magna Carta. In 1219 AD the Mongols utterly laid waste all of the Persian Empire of Khwarezm and in 1241 AD these Asian marauders invaded Poland and Hungary. In 1254 AD Pope Innocent IV authorized torture as a method by which the Inquisition could use in extracting a "confession" of heresy and in 1258 AD the Mongols crushed the Abbasid Dynasty in Baghdad. In 1260 AD several comet groups were born by a close pass of NIBIRU (see *Anunnaki Homeworld*) and in 1268 AD a quake killed 68,000 at Cilicia in Asia Minor.

This is only *half* of the 414 years that began in 1062 AD, with the Turkish invasion of Greece. In 1274 AD Kublai Khan, grandson of the infamous Mongol ruler Genghis Khan, tried to invade Japan with an expansive naval force that was drowned in a typhoon. Seven years later in 1281 AD he again attempted to invade Japan with an even larger navy, and again, a typhoon buried them in watery graves. The Habsburg Dynasty of Austria began in 1276 AD, one of the ruling elite families of the world, reigning through its global bank holdings. This dynastic timeline began a *476 year* countdown to the 1952 AD establishment of the Bilderberg Group, a collective of secret elitists allied together under Prince Bernard, a descendant of the Habsburg royal bloodline, for the founding of a World Government, a virtual revived *Roman Empire* (which fell in *476* AD). Major earthquakes resulted in the deaths of millions around the world in 1290, 1293, and from 1298-1314 AD. And from 1298-1314 AD were also documented many comets, falling stars, plague mists and other strange astronomical phenomena. In 1314 AD mankind actually witnessed with his own eyes a great blackness eclipse the stars, as the gigantic Anunnaki planet NIBIRU passed through the inner system (see *Anunnaki Homeworld*). It would not re-enter our inner system and be seen again for 732 years.

More quakes, comets and meteorite impacts occurred in 1333 AD and the Hundred Years War began between England and France in 1337 AD. In 1346 AD the Black Death plague began to decrease the world's population, having first started in Asia, spreading through the Middle East and Africa, Asia Minor and the coasts of the Mediterranean, then into Europe and perhaps even in America. In 1353 AD the Byzantines (Graeco-Roman Christians of Constantinople, previously destroyed by the Turks in 1062 AD) now acquired the military assistance of their former Turkish enemies in their struggle against the Serbian Kings. The quake of 1356 AD damaged the Great Pyramid and Muslim engineers began extracting the white polished limestone casing blocks for rebuilding materials, even erecting the famous Mosque of Sultan-Hassan in Cairo. In 1362 AD (300 years after this Cursed Earth period began tolling) a Norse-Goth expedition sent by King Magnus of Norway explored North America 130 years prior to Columbus' discovery of the American islands in 1492 AD. This *130 years* is a *tenth* of the 1300 years from the Fall of Rome in 476 AD to the 1776 AD start of the United States of America, founded by *descendants of Norse and Gothic-peoples*, among others of Germanic ancestry. These explorers from Europe left behind the *Kensington Runestone*, which was found in Minnesota in 1898 AD. This stone recorded the account of their exploration of North America and a battle between them and Indians that resulted in some of their kinsmen's deaths. This record and expedition of 1362 AD was precisely *414 years* before the 1776 AD beginning of the United States.

In 1368 AD the Mongols were ousted from China and in 1400 AD the Burning Times were begun, when European governments and the Papacy conspired to confiscate the wealth of tens of thousands of widows left behind by fallen soldiers of the many wars and Crusades. The Church labeled them as heretics and witches, tortured them until they confessed to imaginary crimes and then burned them at the stake before the people. Once dead, the state and church split their spoils. Joan of Arc was burned alive in the Hundred Years War in 1431 AD in the same year Vlad Dracul was born, later to become *Dracula,* and Vlad the Impaler. His life is integral to understanding how this Cursed Earth period unfolds, which began in 1062 AD. During Vlad's life the printing press was invented by Coster of Harlem in 1440 AD, a Dutchman, the victim of a theft by his assistant who took the invention to the one man now believed by largely everyone to have invented it: Gutenberg, who actually only improved upon it. In 1453 AD the Turks captured Constantinople exactly 100 years after the Byzantines had invited them into Europe in their war against the kings of Serbia. Constantinople was named after the Roman Emperor Constantine. As Rome's first and last ruler was named Romulus, Constantinople's first and last ruler was named *Constantine,* the final being Constantine *XIII.* It is interesting to note that Rome fell in 476 AD, and that *476 years* after the Fall of Constantinople was the year 1929 AD, when Vatican City in *Rome* became an independent sovereign state *within* Italy. In this same year of 1453 AD, when Constantinople fell, Gutenberg printed his first Bible and the Hundred Years War (lasting 116 years) ended. Vlad was alive in these years and remained a constant vex to the Turks.

In 1456 AD a bizarre comet with two tails appeared at the time that Vlad Dracul was ruling in Transylvania (Wallachia). He was made infamous throughout Europe and in the Turkish world for his habit of slowly impaling people, transfixing men and women upon stakes and erecting them in a field before his castle in a fate more painful and slower than crucifixion. In 1459 AD Vlad the Impaler killed 30,000 merchants and others in Brasov, Transylvania by impalement on St. Bartholomew's Day. His wicked traits inspired both the Blood Countess Elizabeth Bathory, who bathed in the blood of hundreds of young girls she emulated and Ivan the Terrible of Russia, who nailed people's hats to their heads. In 1461 AD Mohammed the Conqueror set out to kill Vlad Dracul for his successful exploits against the Muslim invaders and his impaling of all Islamic captives. When the Turks reached Vlad's castle in Tirgoviste they were repulsed by the sickening sight of over 20,000 people erected high in the air upon stakes that penetrated them through their private parts, many of them still alive and moaning in a hellish chorus of agony. Dracula escaped but in 1476 AD, exactly *414 years* from the Turkish invasion of 1062 AD, the Turks captured what they considered was Enemy Number One, Vlad the Impaler, of the Order of the Dragon. He was decapitated and his head prominently displayed, impaled on a stake at Constantinople. All of these events and more occurred during this single 414 year span of human history, beginning and ending with the Turks. For the full accounts of these events see *Chronicon.*

We have seen that the Cursed Earth system was originated by the mechanics of astronomical phenomena. The 414 years relates to the lesser periods of 138 years, the actual orbital period of planet Phoenix (138 x 3). The 414 years is found also in other parallels. A comet was recorded by Europeans to have been visible in 1348 AD, the same year a pestilential wind affected Cyprus and many died of asphyxiation during an earthquake. (9) By this time millions of Europeans had died of the Black Plague. Now, 414 years later in 1762 AD astronomers reported that black, unknown objects passed between the sun's surface and Earth's position. (10) Interestingly, 1762 AD was exactly 4000 years after the Great Flood in 2239 BC (1656 AM). In the year 1490 AD a meteorite shower rained upon Central Asia killing 10,000 people, while on the other side of the planet in Ireland at the Ox Mountains, a volcano erupted killing livestock and people. (11) This was 414 years before 1904 AD, when a meteorite collided into the earth near Alta, Norway and a fungal blight killed virtually every chestnut tree of America. (12)

In the previous chapter it was shown how five Cursed Earth periods was 2070 years, and now, it is to be noted how ten Cursed Earth periods, or 4140 years, is also a significant period in human history. We discovered earlier that Ubarutu, the seventh king of the Anunnaki before the Flood, began his reign in 2291 BC (1604 AM). The Mesopotamian records of his reign and kingdom were preserved upon Sumerian and Babylonian tablet texts. His first regal year before the Deluge began a *4140 year countdown* to 1850 AD (5744 AM), when archaeologists discovered the ancient underground and lost library of Assyrian king Ashurbanipal (renowned for his researches into *pre-flood* texts) at the ruins of Nineveh, which yielded approximately 100,000 cuneiform tablet and cylinder texts concerning Assyrian history, prehistory, politics, regnal periods, astronomical tablets, financial transactions and accounts about the *Seven Kings* before the Deluge.

Of even more mind-numbing coincidence (if ever such a thing exists) is the calendrical fact that the Great Flood in 2239 BC (1656 AM) which ended the reign of the Seventh King, began itself a 4140 year timeline to 1902 AD (5796 AM), the END of the Cursed Earth Chronology of 414 year periods spanning back to 4309 BC, when the Pre-Adamic World was ruined by the rebellion of the Anunnaki. What shocks the conscious is the fact that 1902 AD (*6210 years*: 414 x 15 after Pre-Adamic Earth destroyed) was the year the archaic Babylonian tablets known as the *Enuma Elish* were translated into English from Akkadian. These same texts recorded how a planetary body called KINGU was responsible for the destruction of the Pre-Adamic World (which *began* the Cursed Earth Chronology). KINGU and the Anunnaki are mentioned in these olden writings and we have to wonder if KINGU is not the original name for the planet our predecessors knew as Phoenix, a planet that when visually seen, denoted a change in *KING*-ship. The year 1902 AD is integral to our thesis, for so much transpired in this one year that another chapter will have to be devoted to explore it.

Now, having ascertained that the differential between the Phoenix Cycle and the Cursed Earth period produces the 138 year orbit of planet Phoenix, we will now review the history of the world in *138 year Phoenix orbits*.

VII

Orbital Chronology of Planet Phoenix

So it is not only that many eclipses unrecognized by astronomers, as eclipses have occurred, but. . . there are several allusions to intense darknesses *that have occurred upon this earth, quite as eclipses occur, but are not referable to any known eclipsing body.*

—Charles Fort, *Book of the Damned*, 1919 AD

No theory of earth history should be acceptable without a workable chronology. It has been shown that planet Phoenix maintains a 138 year orbit and that this orbit accounts for both the 414 year Cursed Earth periods and the 552 year Phoenix Cycles, as well as providing the chronologist with Year One of the 6000 year timeline of the Curse of Man, and his exile until the return of the Chief Cornerstone in 6000 Annus Mundi (2106 AD) at Armageddon. The Phoenix history spans back 414 years before Year One to a Pre-Adamic cataclysm that ended Earth before Eden.

It must be admitted that several times throughout history when Phoenix entered the inner system, earth was not always in the same place in its journey around the sun, therefore Phoenix did not transit and without a transit it could not occult the sun. Nor could Phoenix be seen in the daytime or when the skies were overcast. If earth was on the opposite side of the sun it could not be viewed, either. Of course there have been sightings of Phoenix lost to history, the records of the ancients not preserved as well as multitudes of legends and lore that are not attached to any certain date. But the following timeline of Phoenix orbits, despite its incompleteness, is sufficient enough to demonstrate that Phoenix exists, was feared by our predecessors, has been seen in modern times and will return with devastating effect very soon. . . in fact, over a billion people on earth today will still be alive to witness and experience this incredible phenomenon.

4309 BC (-414 AM)
Phoenix nearly collides into Earth and the Pre-Adamic World is destroyed.
This is one Cursed Earth period before Year One Annus Mundi (3895 BC) and
2070 years (414 x 5) before the Great Flood of 2239 BC (1656 AM).
138 years (1656 months)

4171 BC (-276 AM)
Phoenix crossed ecliptic in inner system; no records.
138 years

4033 BC (-138 AM)
Phoenix crossed ecliptic in inner system; no records.
138 years

3895 BC (1 AM)
Phoenix seen as *fiery flaming sword* in Year One of Man's Banishment from
Eden, starting 6000 year timeline until man's Redemption in 2106 AD. This
is 1656 years (414 x 4) before the Flood in 2239 BC (1656 AM).
138 years

3757 BC (138 AM)
Phoenix crossed ecliptic in inner system; no records.
138 years

3619 BC (276 AM)
Phoenix crossed ecliptic in inner system; no records.
138 years

3481 BC (414 AM)
Phoenix cross ecliptic in inner system; no records. First Cursed Earth period
of 414 years complete.
138 years

3343 BC (552 AM)
Phoenix crossed ecliptic in inner system; no records. First Phoenix Cycle of
552 years complete.
138 years

3205 BC (690 AM)
Phoenix crossed ecliptic in inner system. This was the 1080th day of Enoch's
reign over the Sethites, renowned chronologist. A multitude of traditions
connect Enoch (Pa-Hanok) to the story of the Phoenix. Enochian texts assert
that Enoch was learned in all the secrets of astronomy.
138 years

3067 BC (828 AM)
Phoenix crossed ecliptic in inner system. This ends the 140th year of Enoch's reign before the Flood. This is also the completion of two Cursed Earth periods and the exact midpoint between Eden and the Flood.
138 years

2929 BC (966 AM)
Phoenix crossed ecliptic in inner system. This begins a 2070 year (414 x 5) timeline to 859 BC which, in the Peruvian chronology, was the year of the 32nd king of Cuzco. Historian Montesinos recorded that the Peruvian text spanned back 2070 years to begin at this time. In 859 BC the Assyrian King Shalmaneser III ascended the throne who, according to Assyrian reliefs and seals, expected the *return* of Phoenix in that year. Great Pyramid under construction.
138 years

2791 BC (1104 AM)
Phoenix crossed ecliptic in inner system; no records. This is completion of second Phoenix Cycle of 552 years.
138 years

2653 BC (1242 AM)
Phoenix crossed ecliptic in inner system; no records. This is completion of third Cursed Earth period.
138 years

2515 BC (1380 AM)
Phoenix crossed ecliptic in inner system; no records.
138 years

2377 BC (1518 AM)
Phoenix crossed ecliptic in inner system; no records.
138 years

2239 BC (1656 AM)
Phoenix crossed ecliptic seven days before the Deluge and transited, *darkening the sun*. This began the fracturing of Phoenix. Full account of the Flood is given in *Anunnaki Homeworld*. This is completion of three Phoenix Cycles and four Cursed Earth periods from Year One (3895 BC), being 552 x 3 and 414 x 4. This is first time these systems synchronized in history. Flood was 2070 years (414 x 5) after the ruination of the Pre-Adamic World in 4309 BC. This year is the exact midpoint in the Age of the Phoenix, counting 1656 years

back to Eden and 1656 years forward, to the sun-darkening of 583 BC.
138 years

2101 BC (1794 AM)
Phoenix crossed ecliptic in inner system; no records.
138 years

1963 BC (1932 AM)
Phoenix crossed ecliptic in inner system; no records.
138 years

1825 BC (2070 AM)
Phoenix crossed ecliptic in inner system; no direct records. This was 2070 years (414 x 5) after Eden and was Abraham's 12th year translating the Giza Texts from off the surfaces of the Great Pyramid. This project is detailed in *Lost Scriptures of Giza* and *Chronicon*. The Great Pyramid and Sphinx site was built to commemorate mankind's fall from Paradise and the promise of his redemption by the return of the Chief Cornerstone (the missing Apex stone the pyramid was supposed to support). The Fall transpired in 3895 BC and was immediately followed by the appearance of Phoenix as a fiery flaming sword (ancient description of a cometary body). The Great Pyramid was finished in 2815 BC (1080 AM), requiring 90 years to complete and was started by the Sethites in 2905 BC (990 AM). Exactly *1080 years* later in 1825 BC, Abraham was teaching the learned of many civilizations the knowledge of astronomy preserved by the ancients, even information on *Phoenix*. This may be because Phoenix could have been seen distantly in this year.
138 years

1687 BC (2208 AM)
Phoenix crossed the ecliptic and transited, *darkening the sun* and causing global earthquakes that felled many megalithic cities, scattering the five armies of Canaan and Amorites that thought to kill off Jacob and his sons after Shechem raped their sister Dinah. Stonehenge III was conducted after the quaking damaged the site. This is the completion of the fourth Phoenix Cycle of 552 years from Eden in 3895 BC, one Phoenix Cycle after the Deluge in 2239 BC.
138 years

1549 BC (2346 AM)
Phoenix crossed ecliptic in inner system; no records. Previously, fractured regions of the wandering planet's frozen surface broke away at perihelion into a comet group.
138 years

1411 BC (2484 AM)

Phoenix crossed ecliptic too close to the sun or during heightened solar flare activity and glacier-to-mountain sized fractures broke off the surface of the dead planet, creating the *Joshua Comet Group* in 1549 BC. Now, as Phoenix passed through inner system, it had a hundreds-of-millions-of-miles long detritus train four years behind it that entered the inner system in 1407 BC and provided the background for one of the most unusual biblical stories, as will be seen in the chapter on the *Joshua Comet Group*. This year of 1411 BC is 2484 years (414 x 6) Annus Mundi, or from Eden in 3895 BC, the 2484 year period noted by the astronomer Aristarchus as the *cycle for cataclysm*.
138 years

1273 BC (2622 AM)

Phoenix crossed ecliptic in inner system but did not transit. Its appearance in the night skies provoked the Assyrians to adopt the Winged Disk as their Seal and this year began the Assyrian Empire. Assyria annexed the famously ancient city and territories of Babylon.
138 years

1135 BC (2760 AM)

Phoenix crossed ecliptic in inner system and transited, *darkening the sun* during the reign of Nebuchadnezzar I of Babylon, whose annals recorded that in this year the sun was darkened by a comet. The appearance of Phoenix prompted a change of kingship over Assyria, for this was the year Tiglath-Pileser I ascended the throne. Olden Irish texts reveal that during the Battle of Magh-Tureadh fought in this year between the Danaan and the Fomorii giants, the sun darkened and the Chinese left records of the darkening of the sun as well. This completes the fifth Phoenix Cycle of 552 years from Eden, two cycles after the Flood.
138 years

997 BC (2898 AM)

Phoenix cross ecliptic in inner system; no records.
138 years.

859 BC (3036 AM)

Phoenix crossed ecliptic in inner system and was seen by the Assyrians who were *expecting it*. This was 414 years after the Assyrian Empire began in 1273 BC, and Babylon was annexed. In this year King Shalmaneser III ascended the throne of Assyria, a king made famous by the biblical account of King Jehu paying him tribute, a fact now proven on the Black Obelisk of Shalmanser. On the obelisk above Jehu's bowing form is the seal of Phoenix. This was also the 42nd reign of the kings of Cuzco, a date according to the Peruvian records that was *2070 years* after the beginning of the Andean Chronology, which we

find in this timeline was 2929 BC, the passage of Phoenix through the inner system before the Flood.
138 years

721 BC (3174 AM)
Phoenix crossed ecliptic in inner system; no records. This was the year the remaining Tribes of Israel were conquered by the Assyrians under Sargon II and deported into Assyrian frontiers, where they assimilated with the Germanic tribes under Assyrian dominion. This is 138 years before the end of the Age of the Phoenix.
138 years

583 BC (3312 AM)
Phoenix crossed ecliptic and transited, *darkening the sun* during a battle between the Medes and Lydians, which was specifically predicted by Thales of Miletus. This is 1656 years (414 x 4) after the Flood and ends the Age of the Phoenix. Planet Phoenix would not begin to synchronize into transit alignment again until the Last Days, with the Cursed Earth system ending *2484 years* (414 x 6) from this date to 1902 AD (5796 AM).
138 years

445 BC (3450 AM)
Phoenix crossed ecliptic in inner system; no records.
138 years

307 BC (3588 AM)
Phoenix crossed ecliptic in inner system; no records.
138 years

169 BC (3726 AM)
Phoenix crossed ecliptic in inner system; no records. This is the final year of peace in Judea, for the following year Antiochus Epiphanes IV of Syria will attempt to eradicate the Jewish religion. He pollutes the Temple with pig's blood and holds public executions of Jews who refuse to Grecianize. This is also the last year of Macedonian autonomy, for Rome will dismantle the kingdom in 168 BC. This completes nine Cursed Earth periods from Eden, and is 2070 years after the Flood.
138 years

31 BC (3864 AM)
Phoenix crossed ecliptic but not in direct transit, causing earthquakes during the Battle of Actium that resulted in 30,000 killed and the ruination of many settlements in Judea. This is 552 years after the Age of the Phoenix ended in 583 BC, when the sun darkened. Had the sun darkened in this year the Legend

of the Phoenix would have lived on. This year completes the seventh Phoenix Cycle from Eden.
138 years.

108 AD (4002 AM)
Phoenix crossed ecliptic in inner system; no records.
138 years.

246 AD (4140 AM)
Phoenix crossed ecliptic in inner system; no records. The Gothic descendants of the Ten Tribes of Israel deported into Assyria and assimilated with the ancient Germans, who migrated into the Roman provincial territory of Dacia and took possession of it, assimilating the Dacians. This is completion of the tenth Cursed Earth period.
138 years

384 AD (4278 AM)
Phoenix crossed ecliptic in inner system; no records.
138 years

522 AD (4416 AM)
Phoenix crossed ecliptic in inner system unseen from earth at *same time* that the Anunnaki planet NIBIRU passed through the inner system after 732 years, ascending from the nether regions of the solar system off the ecliptic. This is the ONLY year in human history that both Phoenix and NIBIRU visit the inner system together. The full orbital chronology and history of NIBIRU is provided in *Anunnaki Homeworld*. This completes the eighth Phoenix Cycle from Eden.
138 years

660 AD (4554 AM)
Phoenix crossed ecliptic in inner system; no records. This completes the eleventh Cursed Earth period from Eden. Further fracturing of Phoenix's frozen liquid surface creates another strewn field at perihelion, spread out in a seven-year train of debris called the *Vials of Phoenix Comet Group*. Their orbital chronology will be covered later in this book.
138 years

798 AD (4692 AM)
Phoenix crossed ecliptic in inner system; no records.
138 years

936 AD (4830 AM)
Phoenix crossed ecliptic in inner system; no records.
138 years

1074 AD (4968 AM)
Phoenix crossed ecliptic in inner system; no records. This is completion of ninth Phoenix Cycle from Eden (552 x 9) as well as twelve Cursed Earth periods.
138 years

1212 AD (5106 AM)
Phoenix crossed ecliptic in inner system; no records.
138 years

1350 AD (5244 AM)
Phoenix crossed ecliptic in inner system; no records.
138 years

1488 AD (5382 AM)
Phoenix crossed ecliptic in inner system; no records. This is completion of thirteenth Cursed Earth period from Eden.
138 years

1626 AD (5520 AM)
Phoenix crossed ecliptic in inner system; no records. This is the 10th and final Phoenix Cycle of 552 years (5520 AM is 552 x 10). Phoenix Cycle system *ends* because another 552 year period cannot fit into the 6000 year timeline, for there remains only 480 years left. In this year of 1626 AD, New Amsterdam was founded after the area was purchased from the Manhatta Indians by the Dutch. This city would later become *New York City*, the financial epicenter of global economy and seat of the United Nations. Therefore, New York was founded in the final year of the Phoenix System. Remember this fact.
138 years

1764 AM (5658 AM)
Phoenix crossed ecliptic in inner system in a partial transit in May. It was seen by many to the naked eye as it covered one-fifth of the sun's surface. Hoffman of the Royal Astronomical Society in England reported that he viewed it through a telescope. He concluded that it was a gigantic object in space that partially passed over the surface of the sun as seen from earth, moving on a *north-to-south* trajectory passing *over* the ecliptic. (1) This is the *first* sighting by modern scientists. That a huge fragment of Phoenix has already detached and spread out along an immense train is revealed by the fact that in 1762 AD (August 9th) a massive spindle-shaped body was seen through telescopes to pass over the sun's surface and did not disappear until September 7th. Its blurred perimeter revealed that it still had frozen liquids burning off of it at perihelion. It was viewed by M. Rostan of Basle, France and by M. Croste in Germany at Sole. (2) More asteroid-comet debris was

seen in 1763 AD and 1764. (3) The fact that these astronomers saw this object in 1762, located on different areas of the sun, reveals that these fragments are closer to the earth than the sun. This is demonstrated in 1763 AD when a large object cast a shadow of *black darkness* on London, England on August 19th, 1763. The appearance of Phoenix, *on schedule*, unveils clearly that this is a planetary-sized object with a *fixed orbit* and, like comets, planets can orbit the sun vertical to the ecliptic. In fact, as shown in *Anunnaki Homeworld*, the planet NIBIRU also traverses the ecliptic in this same way. Phoenix was anciently responsible for the volcanic activity of Mount Vesuvius that buried the settlements of Italy in 1687 BC. This same volcano later entombed Pompeii and Herculaneum in the same manner, so we are not surprised to learn that in 1764 AD Johann Joachim Winchelmann published his *History of Ancient Art*, which covers the secret excavations that were being conducted at the sites of the Roman cities of Herculaneum and Pompeii. (5) Many of the relics and artwork found in these buried ruins are still withheld from public view because of their sexually explicit nature. In the train of Phoenix, four years in following, smaller fragments entered earth's atmosphere. A particularly large one in 1768 AD illuminated North America and was named "Panther Passing Across" by the Shawnee people, the same name given to a baby born that exact night, who would later be famously known as *Tecumseh*. (6) As Phoenix appears to be breaking apart, this is also what we will find 138 years later.

138 years

1902 AD (5796 AM)

This year is the END of the Cursed Earth system of 414 year periods that began with the destruction of Pre-Adamic Earth in 4309 BC, this being 6210 years (414 x 15). This year is 2484 years (414 x 6) after the end of the Age of the Phoenix, which ended in 583 BC (3312 AM) when the sun darkened, predicted by Thales. This year is 4140 years (414 x 10) after the Deluge in 2239 BC (1656 AM). Prior to this date the Great Seal of the United States was that of the Great Pyramid and the *Phoenix*, and now, in this year, the U.S. government quietly changed the Phoenix to the *Eagle*. This parallels the historical fact that Rome's earliest symbol was that of the Phoenix, only later changing it to the Eagle. (7) The symbol seems to be attached to the concept of *empire*: the United States, Rome, Assyria, even the ancient Hittites had a Double-Headed Phoenix. (8) The timing of the change from Phoenix to Eagle hints that there are those in America who knew the significance of the fateful year of 1902 AD. What occurred was not so much an ending, but a startling *beginning*. Our review of 1902 AD actually starts with 1901.

In 1901 the Code of Hammurabi was excavated, an expansive text of Mesopotamian legislation commemorating the *Anunnaki*, an artifact about 38 centuries old. Also in this year the Cincinnati Astral Society published a list of predictions, among them was one that a gigantic passenger steam liner would sink between Britain and the U.S. with a loss of life because of a shortage of lifeboats. This of course happened in 1912 with the Titanic disaster. Another prediction was that a *new comet* would come into view from the southeastern sky (far away from the ecliptic), a strange comet that actually orbits an *unknown planet*, and that an epidemic of disease would follow. (9) Those who study the quatrains of Nostradamus assert that the prophet told of tremendous changes that would begin in the world *after* 1901 AD. At this time in Scotland was experienced a minor earthquake. (10)

In 1902, on schedule, Phoenix almost transited, but did not. So close to the sun, like so many times before, this alien world was invisible from Earth. However, a comet appeared as predicted in 1901 and evidence that it orbited Phoenix and not the sun is found in that it *broke apart* and was photographed as it fragmented. (11) The comet was named Morehouse. Its demise was due to immense gravitational stress between its planetary nucleus (Phoenix) and the proximity of the sun. A volcano erupted at Santa Maria, Guatemala, killing 1000 people, and Mount Pelee on the French Caribbean island of Martinique exploded, sending a searing hot wavefront of gas into the city of St. Pierre. It incinerated the buildings and killed 30,000, as well as destroying most of the ships out in the harbor. Only one man survived inside the town, a prisoner in a cellar already condemned to die, and one man outside the town, both extremely burned. (12) This is a modern Sodom and Gomorrah story. Auguste Sylbanis was the prisoner who survived, like Lot of the book of Genesis. St. Pierre was a Catholic city. In March, a Protestant missionary entered St. Pierre named George R. Penny, who went from home to home sharing his faith and trying to sell Bibles. This enraged the local Catholic priests, for the Catholics had long maintained the practice of disallowing Bibles to the common peoples, strictly controlling through Mass what the people learned about the Word. The priest, using the local authorities, threatened Mr. Penny with his life if he did not leave St. Pierre immediately. He complied and the priest gathered up all the Bibles confiscated from Mr. Penny and the people who had bought them and burned them in the town square. On Easter Sunday this same priest tied a Bible to a sacrificed pig and dragged it down the streets of St. Pierre as a message to all that Protestant Bibles were not acceptable. The morning after this was done the top of Mount Pelee began smoking and rumbling and continued until May 9th, when, at 7:52 AM, the top of the volcano exploded, capsizing ships out in the harbor and incinerating the city. (13)

From November 12th, 1902 through February, 1903 hundreds of millions of tons of a peculiar dust rained upon Australia, the Philippines, southern China, the south of England, Belgium, Holland, Germany, Austria, over the Atlantic Ocean, Switzerland, Russia, the Pacific Ocean, even raining *mud* over Tasmania. In England alone it is estimated that 10 million tons of this strange matter fell across the land. This raining of earth has no precedent on this scale.

Scientists of the time tested the dust and determined that it contained 23-26% *organic matter*. In Australia, 50 tons of red mud per square mile rained in many parts. (14) This extraterrestrial dust is the material trapped in the frozen surface of Phoenix, which darkened the sun in ancient times. Phoenix is fracturing badly and falling apart. This is further exhibited on May 20th, 1903 when the Lowell Observatory published that an object was seen on the terminator of Mars, which on May 27th had moved and was determined to be a *dust cloud* in space. (15) On September 21, 1903 (autumnal equinox) an eclipse *darkened the sun* at nine-tenths totality, but as London and all of England was overcast with clouds, the astronomers of the day never explored the matter. All that is known for sure is that this unusual darkness could not have been the moon. (16) This strange shadow appears to have been only over England and maybe parts of the Atlantic, which hints that it was caused by the transit of a locally passing object near earth – part of the detritus train of Phoenix. The author Charles Fort, a meticulous and avid researcher into these unusual astronomical anomalies, in 1919 in his *Book of the Damned*, wrote, "I think, myself, that in 1903, we passed through the remains of a powdered world – left over from an ancient inter-planetary dispute, brooding in space like *red resentment* ever since." (17) The crimson color of the dust is yet another clue as to its link to Phoenix, which is connected to the ancestral traditions of the moon turning red like blood.

The following year, 1904, boasts of more fragmentary evidence of a disintegrating planet. A meteorite collided into Alta, Norway, its surviving weight being 198 lbs. A mysterious blight from unknown origin killed virtually every chestnut tree in America. (18) On April 15th, 1904 an *extreme darkness* was so black that people could not go out in the open. It lasted for 10 minutes over Wimbledon, England on a *clear day*. (19) Then, 221 days later, for about 15 minutes that began at 10 AM, *intense darkness* covered Memphis, Tennessee with many people screaming that it was the end of the world. (20) Neither of these were a result of the moon's transit, but these events and others were caused by closely passing bodies that cast their expansive shadows across the surface of the earth.

Aside from the annals of little known astronomical records, 1902 was a powerfully important year. Further impressing upon us the fact that it was a terminal year (being the *end* of the Cursed Earth Chronology), it was also the year Queen Victoria of England died, her reign long recognized by the world as the height of British international prestige and dominance, a reign of 65 years. England waned, the United States waxed and the newly emerging empire altered its Seal from the Phoenix to the Eagle of Imperial Rome. The Census Bureau was founded in *New York* in this last year of the Cursed Earth Chronology. Also in New York the Fuller Building was erected, the city's first skyscraper at 21 stories. This was, in 1902, the tallest building in the world made by modern men (the Eiffel Tower is a metal construction, not a true building). The Fuller Building is a one-of-a-kind structure, shaped like a vertical triangle, and it still stands proudly as a wedge between Fifth Avenue and Broadway. The structure was built exactly two Phoenix orbits (138+138 years) after New York was founded (originally called New Amsterdam), in 1626 AD.

Also in 1902, Washington D.C. was planned out and the Capitol scheme reworked, employing Masonic designs. This is the *seat* of power in the United States. As relating to ancient history, not only were the seven tablets of the *Enuma Elish* translated into English from Akkadian cuneiform concerning the Anunnaki and the destruction of an earlier civilization before Eden (which began the Cursed Earth system in 4309 BC), but in this year was also discovered the Hypogeum on Malta, a vast subterranean complex abandoned in 1687 BC, when Phoenix darkened the sun and quakes destroyed civilizations around the world.

Here is a stunning example of synchronicity. Sponge divers off the coast of Crete in 1900 discovered the remains of a Greek shipwreck which produced relics from the first century BC. In May of 1902 an archaeologist first noticed a metal wheel in a lump of pressed debris from the shipwreck that later came to be known as the Antikyathera computer. So incredibly complex is this mechanism, with its precision-cut differential geared wheels of various sizes, that it is evident that the technology required to manufacture the metal computer is more sophisticated than the mechanism itself. The computer was made some time after 713 BC, which we know because the device employs a year of 365.25 days instead of the prior 360 days before the near-planetary collision of 713 BC (see *Anunnaki Homeworld*), which pushed Earth 1 degree further into space, altering the length of the solar year by adding 5.25 days. Zecharia Sitchin remarks that the pointer on the machine may reveal its *purpose*. (21) The months are inscribed in Greek and the Antikyathera specialist, author Dr. Derek De Solla Price, calculated that the pointer device indicates the year 586 BC. Sitchin believes the close date is 584 BC. But, as this thesis clearly demonstrates, the ancients had no astronomical reason for marking perpetually these years, but had a vested interest in marking *583 BC*, as we have seen how it ended the Age of the Phoenix. The Greek of Phoenician descent, Thales, may have predicted this sun darkening using this same computer. The science of Aristarchus, who believed that disasters involving the entire world were separated by a period of 2484 years, is explained by the fact that the sun darkening of 583 BC (3312 AM) was exactly 2484 years (414 x 6) to *1902 AD*. . . which finds a fragmenting Phoenix raining debris on earth 138 years before it will cause a totally earth-shattering cataclysm. The Antikyathera computer was so advanced that the user could compute astronomical cycles *forward and backward* through time. This discovery of the Antikyathera computer in 1902 reveals that it is a part of the Divine

Chart: Orbital Chronology of Planet Phoenix

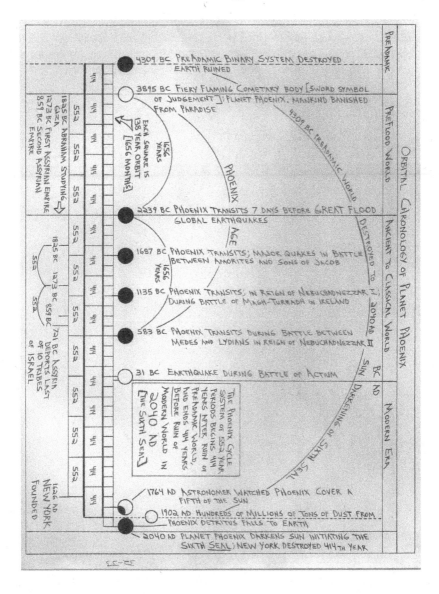

Design, a sign to mankind to search deeper into history to understand where we are in the Divine Chronology.

This last revelation is the key to understanding the purpose of the Great Pyramid in Egypt. As revealed in *Chronotecture: Lost Science of Prophetic Engineering*, the arcane monument is a lithic calendar of the archaic past and future, used to calculate time periods *forward and backward* through history. But the connection 1902 has with the Great Pyramid far transcends this independent fact. This year ends the Cursed Earth Chronology. . .

and what it begins will astonish you.

VIII

Secret Calendar of the Great Seal

of the

United States of America

*Most men, however, find it natural to believe
that lives are predestined from birth, that the
science of prophecy is verified by remarkable
testimonials, ancient and modern . . .*

Tacitus, the *Annals of
Imperial Rome*, VI 21-23

    The founding Fathers of the United States adopted national symbols that have been immortalized upon our country's currency, which has been viewed and valued all around the world. Every people on this earth have seen America's Great Seal, the prominent Eagle clutching 13 Arrows on one side and on the other an image of the Great Pyramid with the Eye cornerstone. Hundreds of millions have perplexed as to the true meaning of these unusual images.

    The final design of the Great Seal was approved by Thomas Jefferson and Benjamin Franklin and its earliest appearance in American symbolism was upon the 1778 AD 50 dollar Colonial Note. The designers of the Seal deliberately published that it was their intention to represent the Great Pyramid in Egypt as the Pyramid, and interestingly, it was acknowledged by the early American government that it was known to them that the original pyramid at Giza was ". . .finished with smooth lines, and that the steps, or indentations now appear, because the surface stones have been removed." (1)

Though the Great Seal was adopted by the Fathers of this nation early on in American history, it still remains a mystery as to why it was concealed from the public eye for 153 years. The Seal of the USA was finally disclosed to the public by President Franklin D. Roosevelt who had it printed on every dollar bill, where it is seen clearly today. This was in 1935 AD. The quiet alteration of the Seal from the Phoenix to the Eagle in 1902 was the esoteric recognition of a change in political status. As with the prior civilizations of Assyria, the Hittites and Rome, that had originally adopted the Winged Disk or images of the Phoenix on their national seals, military standards and coinage, elitists in these cultures understood well that the rise and fall of empires and individual nations were often dependent upon the appearance and cataclysms effected by this strange visitor to the inner solar system. The fall of Phoenix and rise of the Eagle denotes the *end* of a calendrical system, as well as the start of another timeline that sets the perimeters to this end. And the image of the Great Pyramid on the obverse side of the Great Seal provides us with the *identity* of this newer calendar.

In this author's prior work, *Lost Scriptures of Giza*, it is shown that Enoch was the architect of the Great Pyramid, and he was the world's first true Emperor and *chronologist*. He witnessed the future ages of Man and was able to record them in vision and encode them within the three-dimensional spatial confines of the Great Pyramid's chronometry. Enochian traditions are attached to those of the Phoenix and Zecharia Sitchin, in *When Time Began*, demonstrates that even the ancient Egyptians understood that a connection existed between the pyramid concept and that of the Phoenix. (2)   Sitchin has made an astonishing find and published this in his *The Wars of Gods and Men*, where he reveals that an old Assyrian seal has been found that clearly depicted a picture of the two Great Pyramids at Giza with an image of the *Phoenix between them*, called also the Divine Bird of Ninurta. (3)

As the designers of the Great Seal of the USA shared their knowledge that the Great Pyramid no longer had its original casing blocks, we now see in the Providence that moves history how the earthquake of 1356 AD in Egypt that dislodged many of these white limestone blocks was merely a part of the Divine Plan. The Muslim extraction engineers rebuilt Cairo and other villages using these casing blocks, broken up into smaller fragments, especially building many mosques. Had these blocks never been removed, we would have *never* discovered the incredibly simple and gigantic Last Days calendar known as the *Giza Course Countdown*. For 4170 years from the completion of the monument in 2815 BC (1080 AM) to the quake of 1356 AD this 41 story high calendar had remained concealed, and the secret to interpreting it was not unveiled until its 33rd year, in 1935 AD, by publication of the Great Seal. This Last Days chronology began in 1902.

Chart: Giza Course Countdown Calendar

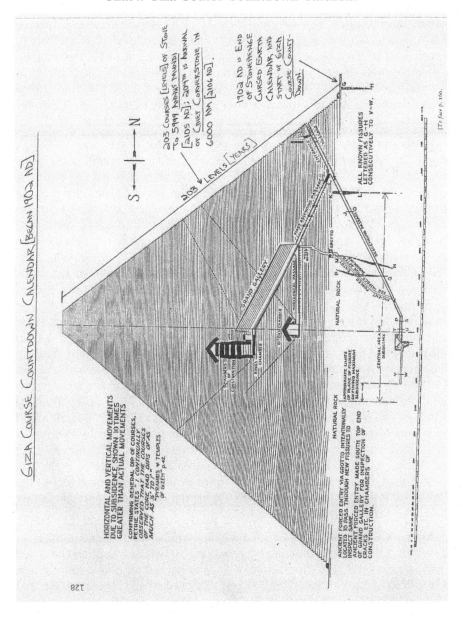

Most universal and divine truths are very simple and are overlooked due to their lack of complexity. And the Giza Course Countdown is no exception. The Great Pyramid of Giza was built by the preflood Sethites at the instructions of Enoch, and this penetrating history is already provided in this author's previous work. The colossal edifice is 203 courses high, or 203 levels of stone blocks. (4) Because all architectural projects begin with a foundation and work is conducted *upwards*, the beginning of this timeline is with the bottom level of brick, representing 1902 AD. The entire structure is layered in virtual sheets of gigantic stone blocks, and every level of this masonry *represents a single year of time*. Each course. beginning at 1902 (foundation), is a single orbit of Earth around the sun. The Cursed Earth Chronology of 414 year periods ending in 1902 AD *begins* the Giza Course Countdown; the Great Pyramid as it stands today, devoid of its casing blocks, is our Last Day countdown of *203 years* until the return of the Chief Cornerstone.

If we add these 203 years to 1902 AD we arrive at the year 2105 AD, which, by Providence, is the 5999th year of the Annus Mundi 6000 years since mankind's banishment from Eden in 3895 BC. This is only *one year* before the year 6000. But the Great Pyramid was *never* finished, its summit terminating in a 20 x 20 foot area, a platform at the apex designed to support a truly gigantic capstone, a fact artistically alluded to in the Eye symbol on the Great Seal. Only one element is missing to complete the timeline and deliver mankind from the *curse* – this being the descent from heaven of the *Chief Cornerstone*, the true Messiah, called the Christ (Anointed One) and the Stone of Israel, or Stone Uncut by Human Hands Who is the Chief of the Corner, all titles given to the Savior throughout the biblical records. The Great Pyramid is a prophetic countdown of stone blocks in layers, each stone representing the redeemed among the history of humanity that form a solid spiritual foundation for the return of the One they exercised their faith in. The structure is the dimensional embodiment of the *Stone Kingdom* wherein the Messiah will begin His reign by crushing the Anunnaki (Fallen Angels) and their human agents at the Battle of Armageddon in 2106 AD (6000 AM) to initiate His Millennial Rule. This Stone Kingdom will bring low all Metal Kingdoms forged by men that have exercised dominion over the peoples of earth, these being Babylonians, Assyrians, Hittites, Persians, Romans, Greeks and a host of others. The Stone the Builders (Anunnaki) Rejected is the *204 level* of the Great Pyramid, showing us once again that the end refers to the beginning. For Christ is the *Word* of God, which was used in the beginning and He will appear on Earth in the End.

Just as this timeline was concealed within the masonry of the monument, so too within a particularly olden apocryphal text known as 2 Esdras (Greek for Ezra), we read that five supernatural beings bestowed all divine information to mankind within ". . .two hundred and four books." (5) These are the 204 levels of the pyramid's masonry. The books here are symbolic of *containers of knowledge*, and as readers of *Lost Scriptures of Giza* will know, there are many references made about the Great Pyramid in distant antiquity, it having been regarded as the largest *book in the world*. . . an architectural library concealing a tremendous amount of information, as also demonstrated in this author's work entitled *Chronotecture: Lost Science of Prophetic Engineering*. Ezra's writing hints that he was aware of this symbolism by the fact that he has *five* supernatural beings convey these knowledges to man, the number five being the geometrical sum of a three-dimensional pyramid's terminations:

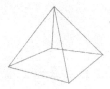

There are many who would by now scoff, declaring that this is preposterous, that this author has merely taken the fact that the pyramid's number of courses is 203 and applied this fact arbitrarily to a year in history, in this case 1902. He has merely discovered in things only that which he put into them, (6) as Frederick Nietzsche would say. But this argument would by necessity have to ignore many corroborating facts carefully outlined throughout this book.

Principle among them are all the sources and facts showing that 3895 BC was Year One of the 6000 year timeline, which would make 1902 AD (5796 AM) exactly 204 years before the 6000 years ends in 2106 AD. Further, the evidence of the 552 year Phoenix Cycle of sun darkenings as well as the Cursed Earth system of 414 years and the 138 year orbital period of planet Phoenix would all have to be ignored to conclude that 1902 AD did not end and begin a timeline – these chronologies all being supported on the authority of ancient and modern astronomical phenomena and findings. Too many coincidences exhibit no coincidence at all.

Further proof that 2106 AD is the year 6000 on the archaic system is found in this author's other works. In *Anunnaki Homeworld* the year 6000 is revealed perfectly by a dimensional analysis of Stonehenge and modern crop formations, as well as the fully detailed orbital history and future of the Anunnaki planet NIBIRU. In *Descent of the Seven Kings* is shown ample charts of historical chronologies involving a 6000 year period and its termination in 2106 AD. In *Chronotecture*, the 6000 years is clearly marked out for everyone to see in the dimensional measurements of the interior arrangements of the Great Pyramid, and in *Chronicon: Timelines of the Ancient Future*, the reader will find an immense timeline of the history of the world, including the 6000 year

Chart: Cursed Earth Chronology

MILLENNIAL

1902 AD IS 204 YEARS TO 2106 AD ARMAGEDDON (6000 AM)

1902 AD IS FIRST COURSE OF BRICK

6000 AM (2106 AD) CHIEF CORNERSTONE DESCENDS FOR ARMAGEDDON

MODERN ERA

2046 AD 144TH LEVEL OF BRICKS.

2046 AD COMET OR COMET OF NIBIRU DARKENS SUN BEFORE COMET IMPACTS N. AMERICA.

1902 AD [5796 AM]

2484 YEARS OF ARISTARCHUS IS SIX CURSED EARTH PERIODS

AD   BC   AD   BC   AD   BRESHEARS '07

ANCIENT-CLASSICAL

CURSED EARTH CHRONOLOGY OR STONEHENGE CHRONOLITHIC

SYSTEM ENDS AT 1902 AD, MARKING YEAR, ONE OF THE GIZA COURSE COUNT-DOWN CHRONOLOGY, [EACH LEVEL OF BRICK IN GREAT PYRAMID IS A YEAR IN THE LAST DAYS]

583 BC [3318 AM] COMET DARKENS SUN STOPPING MEDO-LYDIAN WAR

1825 BC ABRAHAM IN EGYPT

583 BC

2239-583 BC 1656 YEARS

ANTEDILUVIAN

COMET TRANSITS DARKENING SUN BEFORE IMPACT.

2239 BC

(GREAT FLOOD 2239 BC)

2070 YRS.

2070 YEARS

3895 BC MAN BANISHED FROM EDEN

3895-2239 BC 1656 YEARS

144 YRS OF EDEN 4039 BC

PRE-ADAMIC

4309 BC COMET IMPACT AND SUN DARKENS

4309 BC

5239 BC

chronology within a synthesis of over 40 ancient and contemporary dating systems, even chronicling the *exact years* in the future for all the events of the Apocalypse found in the Book of Revelation.

Though we can lead one to water we cannot make him drink. Tacitus remarked that there are those among men who could not even by disasters be brought to the truth. (7) Our sagacious poet Lucretius wrote about this mindset in his *On the Nature of the Universe*. "...since you could see no truth in anything before, how do you know what is to know, and what again *not* to know? What gave you the idea of true or false? What proves to you that there's a difference, that the doubtful and the certain are not the same?" (8) And Pliny, of course, adds, "How many things are judged *impossible* before they happen?" (9)

Our teachers today assume too much, believing that in their short lifetimes in the present they have discovered fact from fictions they perceive from the records passed down from distant millennia. Though our predecessors left behind standing monuments, we cannot replicate many of them today, even with the benefits of modern technology and engineering. In addition, many ancient mysteries remain virtually unsolvable while employing today's interpretative standards, which are too restrictive, so we are steadily brainwashed into believing that the accounts these people left behind for future discovery are not accurate reflections of ancient happenings or ideals. This prejudicial mindset is in direct opposition to all logic and reason because these records are, in fact, *universal*.

We *owe* those who came before us everything we hold true today. Plato wrote in his *Symposium* that knowledge of any given subject is actually a recollection after a departure of knowledge, which is ever being forgotten and later remembered by succeeding generations. He wrote that because knowledge is a *recollection*, this "...necessarily implies a previous time in which we have learned that which we now recollect." (10) Enter Pliny, Lucretius, Thales, Eratosthenes and Aristarchus, stewards of knowledge that had been preserved from even more remote sciences and observations that had collapsed long before these researchers penetrated these mysteries. Plato's own beliefs were partially due to the teachings of his mentor Socrates, who himself taught that his own words were but echoes of more ancient things. (11)

This position was also held by the antiquarian Albert Churchward who, in 1920 wrote, "Thousands of years ago the Grand Architect of the Universe gave the Old Wise Men of Egypt the written laws for life everlasting, as well as *doctrine of final things*. These the human race has never been permitted to lose, and the foundation of our brotherhood was built upon these solid rocks (truths). Although we have traveled far and long since our *originals were in existence*, and we therefore have many innovations in our Rituals on account of compilers of latter days not understanding the originals, the *substance has never been lost.*" (12)

This notion is supported by Maurice Maeterlink who in 1922 wrote, "...the more ancient the texts, the purer, the more awe-inspiring are the doctrines which they reveal." (13) According to the *Book of Sirach*, an old text of uncertain age rediscovered by a Rabbi named Sirach and later included in the Old Testament Apocrypha, the end of the world was recorded and written down in the beginning. (14) This was shown in *Lost Scriptures*, that the Revelation record was remembered under many variant forms by cultures around the world, John's version in the New Testament being a reemergence of this arcane knowledge, previously known and revisited among men in the form of a vision to John.

The original Word of God was known universally but the original religion and its spiritual tenets were fragmented and devolved into many fossilized and impossibly complex pantheistic systems. But the Mysteries and like traditions preserved sacred pieces of the whole; no one people retaining the entire truth. They individually protected jealously many elements of the original ideas while, hidden behind cultural customs, were cleverly masked other pieces of this archaic puzzle, themselves eventually becoming lost to corruptions with the passage of centuries and millennia. But despite all this, from the very outset of history, after this fragmentation of divine knowledge (due to cataclysm), we have contrived to retain our ignorance in order to enjoy an almost inconceivable freedom, according to our philosopher Nietzsche, (15) blindly ignoring that as stewards of knowledge we have a spiritual responsibility to disseminate the information that comes before them. Truth has become disagreeable, so we have allowed fictitious concepts to drag knowledge around like a slave, as Aristotle once put it in his *Ethics*. (16) The modern sage and pyramid researcher Schwaller de Lubicz noted this trend in modern times, noting that the scientific world likes to invest their faith in things stripped of all meaning, while other more meaningful and reasonable explanations are disregarded as fantasy and illusion. (17) Our scientists today are deluded through stringent standards of fact-finding that serve to *omit* information rather than collate it. But this today is no different than that of Lucretius' time, who could have been writing about the scientists of today when he wrote, "What wonder is it then, if the mind misses everything except what it is itself intent on? So from small signs we draw great inferences and lead ourselves into error and delusion." (18) And now that we sneer at those things we do not understand because our modern sages only *pretend* wisdom and knowledge, we are condemned to suffer through a future we were supposed to *remember*.

Because the truth is too impossible for scholarship to grasp, historians adamantly conclude and pass off as fact that the similarities between divergent theologies and civilizations, and their records found continents apart, was simply due to borrowing, migrations and the impossible-to-prove collective unconsciousness of humanity. The truth is merely fantastic and therefore untrue, unacceptable, and any theory advanced that could even be remotely tenable is postulated to the direct contradiction of all known evidence. But while this is the method of the mundane, of intellectuals pretending to assume the throne of gods and of those firmly established in theological dogmas believing themselves allied to God, there are those today who, like the ancient Greeks, do not believe in chance, (19) who hold that every occurrence among the sons of men are a distinct result of mechanics acknowledged to be unknown. It was the ever-searching of the Greek philosophers to uncover the source of these mechanics in the mysteries of nature, phenomena which were recognized as extensions of the attributes of God, for the entire Universe was but a mental image of the Creator.

From the most distant epochs of mankind God has revealed Himself in the mechanics of nature; in the phases of the moon, revolutions of the stars and wandering of planets in the night skies; in forests full of trees, each beckoning men to study them in their perfect symmetry. And in every generation of humanity there have walked those who understood the roar of the tides, the contours of rocks, the worship of insects as they serenade in the wind, the patterns of the stars, passages of comets, the changing of fire and ice and the ever flowing blood of young and old. They clearly understood the world they lived in and the illusions of mortal trappings, which gave them penetrating insight into the world of their ancestors and the *future* glimpses they handed down from the earliest times.

And with this we now look into the future, to see why it was so important to the ancients, to those that followed in Sumeria, Babylonia, Assyria, Hittite Anatolia, Persia, Greece and Rome. . . and to those secret cabalists who in 1902 thought it necessary to remove the Phoenix from the Great Seal of the United States and replace it with the Eagle.

IX

2040 AD Return of Planet Phoenix

It shall come to pass in that day, saith the Lord God, that I will cause the sun to go down at noon, and I will darken the earth on a clear day. . .

—Amos 8:9

The human mind is incapable of manufacturing the synchronicity demonstrated throughout this work of chronology, no singular act of brilliance created this timeline just to compose this book – it was merely by this author *discovered*, not invented. Only a Greater Intellect is able to achieve this level of historical coherency and only because history was written and planned *before* it began. This is noted by the prophet Isaiah who wrote that God ". . .declared the end from the beginning, and from ancient times the things that are not yet done. (1) The Christian apocryphal text of Barnabas reasserts this, reading that ". . .the Lord hath declared unto us, by the prophets, those things which are past; and opened us the beginnings of those things that are to come."

By now it should be evident that the framework of history is through the repetition of cycles involving cataclysms, and that the calendar is found out both backward and forward in time, not by a study of individual years, but by *cycles* of years like the Phoenix Cycle of 552 years or the Cursed Earth periods of 414 years, both having their ultimate origin with an actual *astronomical cycle* that links these together mathematically. The series of historical occurrence had predictive value for those that follow, a fact recognized also by the late Lewis Mumford in his monumental work *Technics and Civilization* wherein he wrote that ". . .the past that is already dead remains present in the future that has yet to be born." (2) This sentiment was popularized also by one of the Founding Fathers of this nation, Patrick Henry, who stated "I have no way of judging the future but by the past." (3)

That the future was never meant to be known is a modern fiction perpetuated by those who don't know it. Knowledge filters have been cast over the truth by our greatest institutions of learning, and history has been concealed by those invested with the responsibility to reveal it. Our learned predecessor Pliny lamented this condition when he remarked that nothing is being added to the sum of knowledge as the result of *original research* and that today's researchers ignore the findings of those that preceded them. Though made almost 2000 years ago, this statement is equally true today. Contemporary writers put out a plethora of books that cover virtually all of the same topics, repeating over and again in different ways what every other author dictates and yet none of these works serve to provide readers with *definite* knowledge of where they live in the timelines operable today.

Although this author is a Christian, he is only one in the original sense. I take serious issue with today's theologians and pulpit prophets who imprison the mind while they act as orators of the highest degree, uttering many fine things without themselves believing or understanding what they say. These modern men of the cloth are learned in all sorts of high matters, their pretense of specific knowledge overshadowing their wisdom (4) and they make themselves masters of cloaking shameful acts in fine words (5) as Herodotus lamented, masking their confusion and lack of knowledge in Latin terms and seemingly advanced doctrines, with the authority of institutions that have financial interests in the Church that supersede the accurate diffusion of truth. Histories found to be offensive are ignored, and for this reason knowledge filters are erected to prevent discoveries that would shatter or compromise prevailing paradigms that currently govern over the minds of men. The attitude of concealing information from the public domain is an archaic trend that antedates even the stone tablet excavated from among the ruins of Nineveh in Assyria (Iraq) once belonging to the scholar-king Ashurbanipal – "The wise shall teach it to the wise; the unlearned shall not see it." (6)

Religion has become the greatest tool for social conditioning and control, alienating people from the knowledge of the future by keeping them chained to the present. Cities rise and fall, nations are plundered, assimilated into other governments, empires are born and fragment, but religions continue through it all. But because the time of the end is a *fixed date*, the mechanics of Apocalypse set in motion long long ago, and because the vast majority of mankind are but slaves to their predispositions and earthly passions, (7) the unveiling of these truths would not make a substantial difference in the outcome of these events upon civilization. However, we must concede that on a *personal* level the acquisition of future information can promote great change. Thomas Burgoyne, Christian mystic and occultist in the 1880s, supports this, having written that knowledge alone is the great liberator of human suffering, our spiritual growth and free will increasing in proportion to our knowledge. (8)

Now, imagine a stone tablet writing preserved in an extinct language from distant antiquity, a relic buried beneath the desert wastes of Iraq (Assyria of old) for over twenty-five *centuries*. This artifact is found and years later it is translated, its content concerning the destruction of the world before our own began, and that the method of ruination was by the close proximity to earth of *objects in space*. When they pass the earth they –

> ". . .bear gloom from city to city,
> a tempest that furiously scours the heavens,
> a *dense cloud* that brings gloom over the sky,
> a rushing wind gust, casting *darkness over*
> *the bright day. . .*" (9)

These texts name the Anunnaki as being responsible for these mysterious objects. The text is from the *Enuma Elish* and as learned previously, these tablets were translated into English in 1902, the last appearance of planet Phoenix in the inner solar system, when extraterrestrial rains of dense *dust clouds* and mud fell by the tons around the world, the year a strange comet appeared and was named Morehouse, an object that orbited Phoenix. This was 138 years after its last appearance in 1764 AD, when Phoenix transited partially and was seen by astronomer Hoffman to have covered one-fifth of the sun's surface. Pieces of Phoenix cast shadows on earth in 1762 and 1763. This appearance was 138 years after Phoenix passed unseen in the year that New York was founded, which happened to be the final year of the entire Phoenix Cycle system of 552 year periods.

The reader was asked earlier in this book to recall two facts: first, was the founding of New York in 1626 AD and the second was that the Phoenix Cycle system that began in 3895 BC with the appearance of Phoenix as a "fiery flaming sword," was antedated by 414 years to the year 4309 BC when the Pre-Adamic World was destroyed, thereby beginning our 138 year Phoenix orbit history. It was this Pre-Adamic ruin that the *Enuma Elish* records recount. In 1902 the Cursed Earth system of 414 year periods ended, but as the past forms a predicate for the future, of necessity do we suffix the Phoenix Cycle timeline that ended in 1626 with a Cursed Earth period of 414 years, just as the system was preceded by such a Cursed Earth period. The Phoenix Cycle of 552 years began and ends with a 414 year period. The date in the future, 414 years after the end of the Phoenix Cycle, is 2040 AD. This forward and backward concept was brought to the attention of the scientific world in the year 1902 AD with the discovery of the Antikyathera computer, and 138 years before 1902 AD was when Hoffman witnessed Phoenix through a telescope and others all over Europe saw it with the naked eye in 1764 AD. 138 years after 1902 is the fateful year of *2040 AD*.

In 1902 AD the Fuller Building was erected in New York, an architectural anomaly and landmark shaped like a triangle in the same year that the Giza Course Countdown begins to count upward through the masonry's 204 levels of stone to the year 6000. The Great Pyramid is a three-dimensional triangle serving as Earth's landmark, and also encodes the precise year of 6000 by its angles. The angular slope of the Great Pyramid's four faces is 51 degrees, and

51 degrees times four faces of the monument is *204 degrees*, mirroring the 204 levels of stone from 1902 to 2106 AD (6000 AM). The pyramid shape has been associated with the tetractys, a series of ten dots forming a triangle. As the ancients believed the monuments represented the sum of ten, we can multiply this ten by the 204 level or degrees to arrive at a calendrical year of *2040* (204 x 10). In our case, this is 2040 Anno Domini.

New York City is a symbol of technological-industrial civilization, of vice and vanity and multiculturalism, founded in 1626 AD, exactly one Cursed Earth period before 2040 AD. The Fuller Building and many other events of 1902 AD count down one Phoenix orbit of 138 years to 2040 AD. Two archaic calendars have ended, both marking through cycles of periodic catastrophes that afflict mankind. All of history is a modular study of the *future*. But in order to authenticate the practical value of the past it is imperative that we *already know the future*. It is not sufficient enough to speculate about what is coming merely based off of what has transpired. This is the domain of prophets. It is through the relationship of spiritual giants who have received knowledge of the future and recorded it for posterity that we can verify the accuracy of cyclic timelines. We have this proof in the form of prophecies such as that written by Amos, that in the distant future God would *darken the earth on a clear day*.

New York City has been the financial, political, cultural and international epicenter of the United States, and by extension the world, a cityscape of gigantic towers like the Fuller Building in the early 20th century as well as the Empire State Building and later World Trade Center Towers. Even the United Nations seat is currently at New York. As the once-mighty empires long ago changed their official seals and standards from the Phoenix to the Eagle before their decline and ultimate end, so too has America followed this precedence. In 2040 AD New York City, at the closure of the Cursed Earth period that began with its founding in 1626, will be totally *destroyed*: not only New York, but many cities around the world will fall in ruins when Phoenix returns to the inner system and nearly collides into Earth:

> And I beheld when he had opened the Sixth *Seal*, and lo, there was a *great earthquake*; and the *sun became black* as sackcloth of hair, and the moon became as blood. And the stars of heaven fell unto the Earth. . . and the heavens departed as a scroll when it is rolled together, and every mountain and island were moved out of their places.
>
> —Revelation 6:12-14

It is the *Seals* of the apocalyptic Book of the Revelation that serve here as the science of prophecy that collaborates this research on planet Phoenix, a study that began with winged disk *seals* on early Mesopotamian cylinders and tablets. Planet Phoenix's appearance in 2040 AD concludes the period of the Seven Seals, the seventh seal being a time of silence and preparation of the Seven Trumpet judgements.

What is disturbing here is that this cataclysmic episode transpires at the *end* of a terribly chaotic period at the start of the Apocalypse, meaning, and quite clearly, that the deceptions, wars, genocide, plagues, starvation, famines and persecutions of true believers that develop in the first five seals have already begun *before* 2040 AD. This is a sobering thought, for the most dreadful period to live through in history is about to begin and humanity as a whole is completely oblivious to this impending global chaos. As it is nearing 2010 AD, we are at the threshold even now, with there being only 30 more years until the *end* of the Seven Seals of the Apocalypse.

In 2040 AD Phoenix will initiate a planetary destruction and poleshift, this final appearance of the fragmenting planet mirroring only its *first* appearance when it resulted in the destruction of the Pre-Adamic World in 4309 BC, precisely one Cursed Earth period of 414 years before man's banishment from Eden in 3895 BC. Thus, this visitation of Phoenix is the colophonic end that refers back to the beginning, just as olden Mesopotamian tablet writings reiterated their final statements in cuneiform, reflecting back to the first statement in the tablet text. Phoenix darkened the sun or affected earth several times in its orbital history, but only the first and last appearance result in cataclysms so horrific.

The return of Phoenix in the Last Days is a theme repeated in the annals of prophecy. In the Book of Enoch we learn that the patriarchal prophet was shown a vision of the future when ". . .the pillar (axis) of earth shook from its foundation, and the sound was heard from the extremities of the earth . . ." (10) About 18 centuries later the Psalmist wrote ". . . therefore we will not fear, though the Earth be removed, and though the mountains be carried into the midst of the sea." (11) To declare such a faith implies that in some remote epoch of human experience the earth had indeed undergone some major changes. That the biblical evidence for planet Phoenix cannot be easily disregarded can be seen in the obscurely dated biblical book of Job. Scholars are at variance over the antiquity of this text, some claiming it is truly ancient, while all conclude it is one of the oldest books in the Bible. The text reads:

> "I would seek unto God, and unto God would
> I commit my cause: which doeth great things
> and unsearchable; marvelous things without

> number: he taketh the wise in their own
> craftiness: and the counsel of the forward is
> carried headlong. They meet with *darkness
> in the daytime*, and grope in the noonday as
> in at night."
>
> —Job 5:8–9, 13–14

Isaiah was a prolific writer on the future Apocalypse, writing that ". . .the stars of heaven and the constellations thereof shall *not give their light*; and the *sun shall be darkened* in his goings forth; and the moon shall not cause her light to shine. . . I will shake the heavens and the earth shall be removed out of her place." (12) He also wrote that the ". . .Lord maketh the earth empty, and maketh it waste, and turneth it *upside down*, and scattereth abroad the inhabitants thereof. . ." (13) The prophet Ezekiel provides us even more detail in confirmation of this thesis. He wrote that God said, "I will cover the heaven and make the stars thereof dark; I will cover the sun with a cloud, and the moon shall not give her light. And all the bright stars over heaven shall I make *dark* over thee." (14) This is no ordinary cloud, but as we were shown in 1902, this is a vast expansive *cosmic dust cloud* derived from Phoenix detritus as the planet continues to fragment.

The prophet Joel not only reconfirms this future event but even provides a clue as to the *timing* – ". . .the earth shall quake before them, the heavens shall tremble: the sun and moon shall be *dark* and the stars shall withdraw their shining. . . and I will show wonders in the heavens and in the earth, blood and fire, and pillars of smoke. The sun shall be turned into darkness and the moon into blood, *before* the great and terrible Day of the Lord come." (15) The Day of the Lord is Armageddon, the return of the Chief Cornerstone, and this is also a *fixed time*, being the year 6000 Annus Mundi (2106 AD). So we see here that the Sixth Seal judgement on human civilization involving planet Phoenix is 66 years before the end. As this is a study on cycles of historic events and their predictive value we find it curious that 66 years *before* 2040 AD was 1974, when the Sears Tower was finished in Chicago, becoming the world's tallest building (again a feat of human architectural engineering), 1974 also being the year that the ancient Chinese Silk Texts were discovered, writings recording many historical disasters. But 1974 was also 66 years after the devastating 1908 explosion of a comet fragment over Tunguska, Siberia that knocked down hundreds of thousands of trees and even people standing 100 miles away. Also in 1908 a terrible quake with a tsunami killed 120,000 people at Messina, Sicily. As the Tunguska comet fell out of the sky, so too, 66 years earlier, in 1842, hailstones rained near Nimes, France that were orange-colored and determined to have contained *nitric acid*. (16) What initiates this amazing 66 year cycle is the start of the United States in 1776 AD, the nation of the Last Days that would show the world its Seal of the Great Pyramid, with the Chief Cornerstone descending upon it as the All-Seeing Eye which would arrive in 2106 AD (6000 AM).

For those seeking a more definite picture of what will occur in 2040 AD when Phoenix nearly collides into our own world, see Appendix A in this book, or read this author's work entitled *Chronicon*, a vast chronology of the world assimilating all the known calendars of old and today serving to put back the pieces of knowledge from around the world so the reader can clearly see the future through the glass of the past. And not just the future, but as Firmicus Maternus noted, even the very beginning of earth and history is detected by the unfolding of historical events that later transpired. (17)

The knowledge of the future was specifically designed by God to be discovered in the end times, the Last Days, for by peering intently into the near future do we more clearly see the past. As is revealed in the Gospel of Thomas, those who can decipher the beginning will know the end. God reveals his greatest mysteries retrospectively, for with the advantage of hindsight can we perceive a more absolute picture of the past operations of Providence at work in human affairs. But the future is not an impenetrable veil, nor a Temple we cannot enter. For there are those among men gifted to see beyond the present fabric of reality, reaching into history not yet attained to grasp secrets from the future they are permitted to bring back to the present. These are they whom understand that a Plan requires a Planner and they comprehend their own place within the Divine Scheme as truly being *timeless*.

The savants of today connected in spirit to the past know fully that knowledge is accumulative and never begins with nothing and is never bequeathed to only one individual. What good is knowledge then if only one man knows? – an old saying, not this author's, but no less true today. We all learn through each other and build our foundation of information upon the discoveries of former students of curiosity and learning. It is our duty to add to this accumulation of knowledge, to take from it that which does not belong and to repair it with accurate information. What we know we *inherit*, for knowledge is all we take with us when leaving the material existence, knowledge is the *only* experience benefiting the human spirit – everything else worked for by man perishes with the body.

The day approaches when every contributor will be given a stone identifying themselves that will be placed within the structure of divinity and in the end there will be left a perfect Temple for God to reside in, an edifice of colossal proportions containing eternal living stones (redeemed humanity) possessing *divine* knowledge, for what we learn in this life and apply will be an inheritance amplified exponentially by God. We are merely leaves in the book of life. When God refashions His tree of elect souls to make His eternal tome, pray that you will be found an adequate building material useful for His everlasting plan.

At the end of this work Appendix B is provided for those seeking a concise review of several legends and myths involving the darkening of the sun, all referring no doubt to the Phoenix phenomenon or comet transits. Because they are merely legends, they are not attached in this study to any of the exact dates previously detailed, though they may all be related. But first, there are many other solaric anomalies and astronomical events to review relative to Phoenix. These are disclosed in the following two chapters. The first concerns the *Joshua Comet Group* and the second is the *Vials of Phoenix Comet Group*, both trains of cometary debris are strewn fields having their origin as glacial fragments that long ago broke free of the surface of Phoenix and entered their own solar orbits.

X

The Joshua Comet Group

He is wise in heart, and mighty in strength:
which hath hardened himself against Him,
and hath prospered? . . .which shaketh the
earth out of her place, and the pillars thereof
tremble. Which commandeth the sun, and it
riseth not: *and sealeth up the stars.*

—Job 9:4-7

As planet Phoenix orbits the sun it suffers entropy, as with everything else in the known material universe. With each pass into the inner system every 138 years (2040 AD being the 45th orbit) at perihelion the dead planet experiences intense gravitational and magnetospheric attraction, tension, friction and stress as it traverses vertically across the ecliptic plane. On occasion frozen sheets tear free from Phoenix's surface which is riddled with fractures that break off into immense glaciers, with dust and rocks strewn within them. These in turn enter into their own orbits around the sun and fragment even more into comets and asteroid trains.

Phoenix is the origin of two such cometary groups. The first group we will review is the *Joshua Comet Group* that, by 1411 BC when Phoenix passed through the inner system, was already four years behind the planet and, upon entering the system close to the sun, this cometary group broke up and away from its gravitational nucleus and entered into its own independent orbit around the sun. It is called the Joshua Comet Group because it begins with an event in the Old Testament involving Joshua and ends with another disaster in 33 AD at the Crucifixion of Christ, named Jesus, a Greek name for the Hebrew *Joshua*.

This comet, a gigantic mountain of ice and rock debris, entered our inner solar system in the year 1407 BC between earth and the sun and was completely invisible as it approached dangerously close to our planet. In Canaan the Israelites had destroyed Jericho and the Amorite garrison-city of Ai, and were met with a vast host of armies led by a multitude of Canaanite and Amorite kings. The Israelites were vastly outnumbered in what amounts to be a repetition of history, mirroring the 1687 BC conflict between the Canaanite-Amorites army and the sons of Jacob (Israel) – when the sun darkened by Phoenix transit. Again here in 1407 BC, a *fragment* of Phoenix causes another global event.

93

God commands Joshua, who is leading the conquest, to utterly destroy the armies before him. As earlier revealed in this work, the curse of Canaan was to be effected in this year because these people, formerly under the curse of Noah for the sexual deviancy of their patriarch, reconfirmed their accursed status by defying the Law of the Lots and taking residence in a land inherited by Shem and his descendants. Joshua complained to God that His request was unreasonable, due to the immense multitudes of Canaanites and their allies assembled for war, and that the Israelites could not possibly slay all of these enemies in a single day. The remedy? The Lord told Joshua to command the sun and moon to *stop moving*. He assured Joshua that He would aid the people in taking the land.

Joshua did as he was told and the sun stopped in the sky and the moon ceased moving over the valley of Ajalon. A storm of stones from the sky rained upon the armies of the Canaanites and confused and decimated their hosts, while the Israelites entered the enemy ranks and cut them down, their duty requiring an additional 10 hours that the day originally could not provide. But the comet from Phoenix penetrated Earth's magnetosphere and entered into a violent gravitational gridlock that visited innumerable lightning blasts (flux tubes) between the earth and the comet, vitrifying entire areas, as quakes trembled the planet. Because the world hangs upon nothing (a fact mentioned in Job 26:7) the axial spin-rate of the planet is caused by its velocity, mass and distance from the sun. This titanic comet only resulted with a cessation of earth's rotation and it was only for a brief period of time before the planet, rid of its intruder, resumed its normal motion. For those wanting to review a more detailed account of the event from the universal eyes of the ancients, see Appendix C.

The year 1407 BC began Israel's occupation of the Land of Promise. The connection to Christ is more penetrating than just Joshua's name (Greek *Jesus*). Both mean *Savior*. The year 1407 BC serves calendrically as an isometric epicenter between two major historical events, both *1407 years apart*, before and after 1407 BC. The Great Pyramid was completed in the year 2815 BC (1080 AM), an architectural calendar, marking events in its internal chronometrical timelines *forward and backward* in time (see *Chronotecture*), a massive edifice serving as a prophetic allusion to the coming Stone Kingdom and the coming of Christ, the Chief Cornerstone of the Monument of Man. The first year after the pyramid complex was complete began a *1407 year* countdown to 1407 BC (when the sun stands still for Joshua). This year itself began a *1407 year* countdown to 1 BC, the birth of Christ, the Stone of Israel, whose name is actually *Joshua*.

While this calendrical fact is interesting, even more fascinating is the orbital chronology of the Joshua Comet Group and how it *ends* with the Crucifixion of Christ in 33 AD. With its tidal struggle with earth in 1407 BC the comet fragmented even more with its own detritus train of fragments. It orbits the sun at precisely 360 years. Every 220 years the group passes under the ecliptic below the inner system and spends likewise 140 years above the ecliptic in far northern space over the inner system.

We shall now review this orbital chronology.

1407 BC (2488 AM)
Near collision with earth; our world ceases to rotate for 10 hours. Colossal comet fragment of Phoenix breaks up. Joshua leads the Conquest of Canaan during disastrous episode.
220 years orbiting below ecliptic

1187 BC (2708 AM)
No records.
140 years orbiting above ecliptic

1047 BC (2848 AM)
No records.
220 years orbiting below ecliptic

827 BC (3068 AM)
No records.
140 Years orbiting above ecliptic

687 BC (3208 AM)
Chinese annals report that on a clear night the fixed stars failed to appear as something in space blocked them out. Toward the middle of the night meteoric rains fell from the sky. (1)
220 years orbiting sun below ecliptic

467 BC (3424 AM)
Paralleling Thales of Miletus in every way, in this year Anaxagoras predicted that in a certain number of days (howbeit Thales only predicted the year) a meteorite would fall from the sky, which he deduced in this the 2nd year of the 78th Olympiad. Like Thales, Anaxagoras too, according to Ammianus Marcellinus, learned how to predict meteorites from the Egyptians. (2) As predicted, a meteorite indeed fell and crashed into Thracia by the river Aegos and that very night a *bright comet appeared*. (3) Another unusual parallel here is the fact that in this same year Aescylus composed his famous work, *The Seven Against Thebes*, which concerned the earliest recorded war or *any* event in known Greek antiquity, a war which transpired in 1244 BC when a comet was seen and recorded by the Assyrians in the Annals of King Shalmaneser I, (4) when it appeared during a battle with the Hittites. By this time the debris train of this rogue comet was extensive, created in 1407 BC, but amplified in

687 BC. This train of large fragments is demonstrated clearly in the records of history. The cometary detritus following in the wake of this gigantic and disintegrating comet appears to be falling behind in a six or seven year train, the smaller fragments losing velocity.

464 BC (3431 AM)
Three years behind the comet during the first regnal year of King Artaxerxes of Persia, (5) Livy reported that in this year the Romans reported strange lights ablaze in the skies, attended by other inexplicable phenomena. So disturbed were the Romans that three days were officially put aside and the entire city virtually closed down, wherein work was forbidden and the people packed the shrines and temples. (6) A destructive earthquake also afflicted Sparta.

463 BC (3432 AM)
Four years behind the comet a plague killed many thousands of Romans as well as much livestock.

461 BC (3434 AM)
Fires blazed in the skies and a violent earthquake shook Rome. It was also reported that it rained lumps of a meat-like substance that birds devoured greedily. It took days before the meat became putrid. So strange was the event that the Senate called for a reading of the Sybilline Books. (7) This was the sixth year behind the comet.

Being so far behind their parent debris train, many of these objects, having lost velocity and mass from tidal forces imposed upon them from the sun and earth, are torn from the comet's trajectory and enter into a tighter yet ellipsoid orbit around the sun as the gigantic comet leaves the inner system. These fragments return 31 years later in 436 BC before vanishing from history. Meanwhile, the main group of fragmenting pieces continues unabated. These lesser fragments reappear in:

436 BC (3459 AM)
Odd and frightening things occurred in the skies, and earth tremors collapsed buildings in Italia. The plague returned with an increased mortality rate. (8) Fragments have spread into an eleven year train.
31 years orbiting the sun

405 BC (3490 AM)
Debris train passes earth unseen; no records. The following year begins disintegration of train.

404 BC (3491 AM)
Athens falls to Sparta at conclusion of Peloponnesian War. An unusual reddish rain fell in this year.

401 BC (3494 AM)
Plague in Rome is being regarded as the Wrath of God. Unusually cold winter followed by record-breaking summer heat. (9)

394 BC (3501 AM)
During the Battle of Cnidus between the Greeks and the Spartans, a meteorite appeared like a beam across the sky as the Spartan fleet was defeated. (10)

As this sums up the orbital history of these fragments, which began breaking off in 1407 BC, as well as 687 BC and then 467 BC, we now return to the *unaltered* orbit of the Joshua Comet Group's huge ice and rock mountain. We left off on this intriguing tangent in 467 BC.

140 years orbiting above ecliptic

327 BC (3568 AM)
No records.
220 years below the ecliptic

107 BC (3788 AM)
No records.
140 years above ecliptic

33 AD (3927 AM)
This is the appearance of the Great Comet at the Crucifixion in 33 AD exactly 1440 years (144 x 10) or 360 x 4 years after 1407 BC, when the sun stood still for Joshua, which initiated this orbital chronology. Jesus is arrested, falsely accused and condemned to death by a Jewish conspiracy in direct contravention to clearly established Levitical law. The Roman governor of Judea, Pontius Pilate, learned from his wife (she had a terrible nightmare) about the supernatural divinity of Jesus, for she begged him not to have anything to do with his death. Her dream specifically portended a dreadful earthquake and *darkening of the sun*. (11) Jesus is delivered to the Romans and is crucified on Golgotha Hill outside the city of Jerusalem to pacify the riotous Jews. A Centurion pierces His side with a spear when an *earthquake shakes the entire world*, splitting the rocks on the hill and warping the walls of the Temple enough that the massive Veil is split in half that guarded the way to the Most Holy Place – an act of God, signifying that sacrifice and ritual was no longer required for atonement of sin, and priests were no longer needed for direct access to God. In the Gospel account, when Jesus died the *sun darkened* for three hours. Many historians of the day remarked that in this year *Phoenix darkened the sun*, (12) clearly associating the darkening of the sun to the archaic legends of a great celestial bird immersing into the sun. The reference to Phoenix is truly astounding, especially in light of the fact that

this comet's origin was from off the surface of Phoenix itself. In this event the gigantic comet transited between the sun and the earth as much of its frozen debris burned off and formed a tail that enveloped the Earth as it traveled over the ecliptic:

> Now from the sixth hour there was *darkness*
> over all the land unto the ninth hour.
> —Matthew 27:45

This was the height of the afternoon, when the sun is brightest in the sky. The quakes continued and shook Syria, Judea and Lower Egypt around the Great Pyramid complex and we read in the *Gospel of Nicodemus* that a terrible darkness covered the earth, (13) and again, it mentions the eclipse to have lasted three hours, quite impossible had it been the moon. In *Pilate's Report to Tiberius Caesar* we learn that ". . .by degrees the day darkened like a winter's twilight." (14) This report to Caesar needs to be more fully examined:

> . . . and when He (Jesus) was hanged (upon the cross) supernatural signs appeared, and in the judgment of the philosophers menaced the whole world with ruin. (15) There was darkness over the whole world and the sun was obscured half a day, and the stars appeared, but no luster was seen in them; and the moon lost its brightness, as though tinged with blood. . . and the terror of the earthquake continued from the sixth hour of the preparation to the ninth hour. (16) . . .the sun was altogether hidden, and the sky appeared dark while it was yet day. . . wherefore, I suppose, your Excellency is not unaware that in all the world they lighted their lamps from the sixth hour unto evening. And the moon, which was like blood, did not shine all night long, although it was at the full. . . So dreadful were the signs that men saw both in the heavens and on the earth that Dionysius the Areopagite is reported to have said, 'Either the Author of Nature is suffering or the Universe is falling apart.' (17)

Pilate's story is one that does not end with the Scriptures. The apocryphal texts detail these reports, and further read, "Now when the letters came to the city of the Romans and were read to Caesar with no few standing there, they were all terrified, because, through the transgression of Pilate, the darkness and earthquake had happened to all the world." (18)   Pilate was then cited to Rome where before the Caesar he declared, ". . .never have I read in the works of the philosophers anything that can compare to the maxims of Jesus." He was then sentenced to die and duly executed, as a believer of Jesus, his wife not suffering to live longer than he. The execution of Pilate was not inconsistent with Tiberius' Caesar's Reign of Terror at that time, for he was, in those years, killing off all political dissidents and forcing many into banishment or suicide.

Rabbi Akiba wrote that it was a dark mist that obscured the sun as the earth quaked, further quoting Celsus, an enemy of the Church who also recorded this disaster. (19)   Roman archives preserved in the Vatican Library by the Church historian Eusebius and the pagan writer Phlegon of Bithynia specifically detail the sun darkening and quake of 33 AD. Interestingly, one of the reasons many historians and scholars refuse to accept the Gospel account of the sun darkening is because computations prove that there occurred at this time *no eclipses*. And, as now evident, through the chronicles of the histories of this book, no eclipses are required. To darken the sun, the debris train from a comet's tail would not have to be much, considering the fact that if Earth had an 8 inch diameter then the sun would only be 70 feet in diameter, but it would be *a mile and a half away*. (20)   That's 7,920 feet apart. A small shadow from an object transiting between the sun and earth would easily secure the darkening effect, especially if the object was closer to Earth than the sun.

In 1628 AD Fernando Montesinos, a Spanish chronicler, spent decades researching the Inca and their history. He composed an extensive regnal history of Cuzco in the Andean mountains of South America. This history was researched by Zecharia Sitchin with some remarkable findings. The Ancient Peruvian Empire, according to Montesinos, maintained a tradition that the sun did not arrive on time, that sunrise was delayed by 20 hours in the 3rd year of the reign of King Titu Yupanqui Pachacuti II, the 15th king of Cuzco. (21)   Amazingly, Sitchin, working only with the data provided by Montesinos, which itself dealt only in approximates, still arrived at a date only *13 years* away from the true year of 1407 BC for this solaric anomaly of the sun standing still.

The parallels between the Joshua Comet Group beginning and ending from 1407 BC to 33 AD, a 1440 year span of time, are also aligned with the ancient Peruvian records. Montesinos recorded that the Inca claimed the *Fourth Sun Age* ended in the reign of the 58th king of Cuzco. This is 43 kings after 1407 BC, King Titu, these having an average regnal period of 33 years, howbeit, it is recorded universally among the nations of long ago that the earliest peoples lived extraordinarily long lives. Moses died in 1407 BC, at the age of 120 years old, giving the leadership over Israel to Joshua. The ancient American civilizations of the Olmecs, Zapotecs, Inca, Maya, Tolteca, Aztecs and others all measured the histories of their ancestors within a framework of *Sun Ages*, believing that when a sun died it was reborn. The Phoenix Cycle of 552 years is what gave rise to such a tradition, so when the sun darkened and the disaster affected the entire world in 33 AD, the Andean scholars among the Inca claimed that the Fourth Age was complete.

We have further confirmation from India and the east of this sun darkening event in 33 AD. The traditions of Crishna antedate Jesus' presence on Earth from 1 BC to 33 AD, Crishna being the physically manifested form of the god Vishnu of the Hindu Trinity. The reason that stories of Crishna and Jesus parallel each other so perfectly (and it is astounding how parallel they are) is because the prophecies of Christ were known to the ancients thousands of years ago but as time elapsed, these future-based texts were corrupted into doctrines that asserted that these events had already been fulfilled by local deities. In the stories of Crishna's crucifixion, a black circle appeared around the moon and the *sun darkened*, as the sky rained fire and ash. Flames burned as terrible calamities transpired over the earth. (22)

This concludes the Joshua Comet Group's orbital longevity and history, a group thematically related to Israel, Joshua and Jesus as well as with the early history of Rome and later, when the Romans crucified Christ and then killed Pilate. If we follow this same orbital pattern of 360 years of the Comet's pieces (220 years under the sun and 140 over) the only other unusual event that occurs is in 1333 AD (5227 AM), when a single meteorite crashed into China killing everyone within 100 miles with noxious gases that filled the air. (23) At this date in 1333 strange things were recorded in the skies over China, earthquakes swallowed entire villages and many cities were ruined. This may in fact be a remnant fragment or the nucleus of the Joshua Comet.

We now review the equally startling *Vials of Phoenix Comet Group*.

XI

Vials of Phoenix Comet Group

And I heard a great voice out of the Temple
saying to the seven angels, Go your ways,
and pour out the Vials of the Wrath of God
upon the earth.

—Revelation 16:1

Some time prior to 660 AD a truly immense sheet of glacial debris broke free of the surface of Phoenix, fragmented and trailed behind the wandering planet in a seven year train. After passing the sun in 522 AD, with Phoenix at perihelion, the train of detritus entered into its own orbit eighteen years behind its parent planet. Approaching the sun, the group tore free from the gravitational influence of Phoenix and initiated its own orbit of 72 years underneath the solar ecliptic and 84 years above it for a total orbital longevity of 156 years. These asteroids and comets orbit the sun and approach much closer to Earth than that of the fixed orbit of Phoenix, often with Earth passing right through their train as it journeys along the ecliptic plane at 20 miles per second.

Here is the seven to eight year chronology of the Vials of Phoenix objects.

678-685 AD
(4574-4581 AM)
No records.
72 years below the ecliptic

750-757 AD
(4644-4651 AM)
No records.
84 years above the ecliptic

834-841 AD
(4728-4735 AM)
840 AD is the approximate date for the appearance of many comets, all within a three or four year period that disturbed Chinese officials. (1)
72 years below the ecliptic

906-913 AD
(4800-4807 AM)
A violent earthquake rocked northern Egypt in the region of the Great Pyramid. (2) The *Irish Annals* record that in 911 AD a comet was seen in the night skies. (3)
84 years above the ecliptic

990-997 AD
(4884-4891 AM)
In 994 AD bizarre omens appeared in the skies and frightened the Toltec people in ancient America. At the city of Tollan they were already agitated by the licentious habits of King Huemac II. That winter a crippling frost destroyed their crops and the following summer was unusually intense. As plague set in, invaders known only as the Cichemic drove them out of North America into the southern areas known today as Mexico. (4) Strange lightning from heaven completely destroyed the Irish town of Armagh, all of its houses, cathedrals, belfries and even the archaic Round Towers that had been standing in the region for almost 2000 years before this date. The *Book of the Four Masters* claims that this was caused by unusual thunderbolts, an event also recorded in the *Annals of Ulster*. (5)
72 years below ecliptic

1062-1069 AD
(4956-4963 AM)
Just weeks prior to the Norman invasion of England a comet was seen. William of Normandy became William the Conqueror, King of England, after defeating King Harold at the Battle of Hastings. (6)
84 years above ecliptic

1146-1153 AD
(5040-5047 AM)
1150 AD is approximate date when many comets and signs in the sky were recorded by the Chinese, who claim that at this time there were ten times as many comets as normal. (7)
72 years below the ecliptic

1218-1225 AD
(5112-5119AM)
No records.
84 years above the ecliptic

1302-1309 AD
(5196-5203 AM)
This is during the famous 16 years of the Seven Comets, a period wherein

were seen and experienced around the world several comets, asteroids passing the earth, meteoric rains, plague fogs descending from the skies, earthquakes and flooding. (8)
72 years below ecliptic

1374-1381 AD
(5268-5275 AM)
No records. By this time the seven year train is spreading out even further into a strewn field of 8, 9 or even 10 years in length as many pieces lose mass and velocity.
84 years over the ecliptic

1458-1465 AD
(5252-5259 AM)
No records.
72 years under the ecliptic

1530-1537 AD
(5424-5431 AM)
The Inca ruler of South America related the ancient prophecy of the coming of the "bearded race," of white men whose arrival would be marked by the Signal of God in the heavens. (9) The Spanish indeed arrived and wrought ruin among the Inca, raped their country of gold and other precious commodities and raped over 5000 of their women. (10) This Signal of God was a comet that appeared in 1530 AD. In 1531 AD Halley's Comet passed and on January 26th a quake rocked Lisbon, Portugal killing 30,000 people. In 1533 AD a bright comet appeared in the sky according to European annals. (11) This is the same year that another group called the Dying Comet Group (from NIBIRU: see *Anunnaki Homeworld*) passed through the inner system. In 1534 AD the Black Death Plague returned to Europe, just as predicted by the young student Nostradamus. (12)
84 years above ecliptic

1614-1621 AD
(5508-5515 AM)
In 1620 in Germany, standing water was reported to have turned red like blood. (13) This reddish contamination seems to be a chief characteristic of the Phoenix-object's chemical makeup.
72 years below ecliptic

1686-1693 AD
(5580-5587 AM)
In 1686 AD a viscous liquid rained in Europe and in Germany a coal-black leafy mass of fragments fell from the sky that was putrid, reported also over Norway and Pomerania and England. (14) In 1687 AD a terrible quake at

Callao, Peru resulted in massive loss of life and property, this year of course paralleling the global quakes caused by Phoenix in 1687 BC. In 1692 an unusually putrid rain afflicted Germany. (15) In 1693 a quake at Catania, Italy killed 60,000 and in 1695 it rained masses of a greasy substance-like butter that was especially foul-smelling over Ireland. (16)
84 years above ecliptic

1770-1777 AD
(5664-5671 AM)

A comet appeared and passed 1.39 million miles from earth, the closest of any in modern history acknowledged by the scientific establishment. (17) The sun is over 90 million miles further than this comet's pass. The comet traveled on and as it neared Jupiter it fragmented and has not been seen since. (18) It was named Comet Lexell. In 1772 a comet appeared, named Biela at the time by astronomers, that then entered into a much tighter orbit around the sun. In this same year a stone fell from the sky at Luce, France. (19) An astronomer in 1777 AD, in June (Named Messier) witnessed through his telescope a vast multitude of dark bodies passing through space. (20) In June of 1779 AD at Boulogne, France, appeared many luminous bodies passing through the air during an earthquake. (21)
72 years below ecliptic

1842-1849 AD
(5736-5743 AM)

One year prior to this cycle (denoting an elongation of the train), in 1841 an oily, reddish matter fell from the sky over Genoa in February (22) and pebbles fell from the sky. even breaking windows in rural area of England. (23) Orange hail fell in 1842 near Nimes, France and was determined to contain nitric acid. (24) An extremely bright comet appeared, having a tail 500,000,000 miles long (sun is only 93 million miles from Earth). (25) In 1844 AD falling stars were witnessed and people explored the regions where they fell, only to discover bizarre grayish jelly-like masses. This was October 8th. On October 4th a luminous object appeared in the heavens that sent out quick flickering waves of light. (26) In 1845 a comet appeared and shining objects passed through the heavens over Naples, Italy. In June three luminous bodies emerged from under the sea near an English brig named *Victoria*, 900 miles east of Adalia, Asia Minor. These UFOs were half a mile from the vessel and were visible for ten minutes until disappearing from the sky. (27) Could it be possible that we are not the only ones interested in Phoenix or these astronomical phenomena? The time of the end and its coming signs would be of great concern to the Anunnaki. Also in 1845 Comet Biela returned to the inner system and struck some unknown dark object late in the year, fragmenting the comet into two large pieces. Biela, in the following year, was seen as two distinct comets, and also in 1846 a formerly unknown comet named Bronson appeared. Its trajectory is so different than other comets that astronomers asserted that Comet Bronson orbited *some other body* unknown to us, and not the sun. (28) Interestingly, in this year planet Neptune was discovered. In April of 1846 a grayish gelatin-like matter rained in nut-sized drops over Wilna, Lithuania and we are informed that the same substance rained over Asia Minor in 1841 AD. (29) In March a weird fibrous material rained over China, and in October, France suffered intense reddish rains. (30) In 1847 astronomers J.R. Hind, Benjamin Scott and Mr. Wray all witnessed through telescopes a very large dark spot pass over the surface of the sun, even naming the object *Vulcan*. The object was also seen by Schmidt and was said to be about the size of Venus. (31) During an eclipse an astronomer named Rankin studied the moon and saw luminous objects upon its darkened surface. (32) On December 8th 1847 the sky blackened on a clear day over Arkansas, USA and strange clouds materialized with a reddish glow that appeared over Forest Hill before a huge explosion was heard in the sky. A meteorite crashed outside of town creating an eight foot hole. Within minutes the sky cleared. (33) During a full lunar eclipse in 1848 the lunar orb appeared red and was not totally darkened, even in the shadow of Earth, as if another source of light illuminated it. (34) A three month long earthquake in the Aegean and Mediterranean regions killed much wildlife and property, though only 50 people died. (35) In 1849 a blackish rain-like ink fell upon Ireland in May and red rains descended upon Sicily and Wales. Ice in irregular chunks (not hail) fell from the sky on Scotland, one piece being 20 ft. in circumference and in November the astronomer Sidebotham saw a gigantic unknown dark object passing through space. (36) In 1850 a black rain fell after an unusually long rumbling sound was heard in the sky over Northampton, England. (37) 84 years above ecliptic

1926-1933 AD
(5820-5827 AM)
In 1927 an earthquake killed 200,000 people in Tsinghai, China. Another quake collapsed the banks of the Jordan River in Palestine disrupting the flow of the river for 21 hours, a phenomenon that occurred in the year 1407 BC when Joshia led the Israelites in the Conquest of Canaan after passing over the Jordan River to the astonishment of the locals. (38) A large meteorite crashed into a remote region of Brazil killing many people and witnessed by a Jesuit missioner, Fray Fidelio, who sent a report to the Vatican. (39) At Jakarta, Indonesia, Mount Merapi erupted. In 1931 flooding in China killed 3,700,000 people.
72 years below ecliptic

1998-2005 AD
(5892-5899 AM)
In 2001 a quake killed 20,000 in Gujarat, India. In early 2003 the Space Shuttle Columbia broke apart upon reentry through the atmosphere raining debris all over Texas. The loss of its heat shield plate or the ruin of the entire shuttle itself could have been due to a solid mass material striking the craft. In 2004 Venus transited and a hitherto unknown asteroid named Apophis (Egyptian god of destruction) was discovered, a near-earth object. The world's most powerful quake in 40 years occurred under the Indian Ocean causing a tsunami that killed about 200,000 people, the quake measuring near Sumatra at 9.2. Some estimate the death toll at closer to 300,000. In 2005 a quake killed 80,000 people in India and Pakistan and hurricanes Katrina and Rita ruined the American coasts along the Gulf of Mexico, Katrina accompanied with 33 tornadoes. The comet train itself may have largely passed on the other side of the sun, earth passing only through its thinnest areas when intersecting the train along the ecliptic.
84 years above ecliptic

2082-2089 AD
(5976-5983 AM)
This is the final time the Vials of Phoenix afflict the Earth, this being the dating for the Vials of the Wrath of God, Seven distinct plagues that will afflict mankind long after the Seven Seals and the Seven Trumpet judgments. During the Beast Empire of the Antichrist (a man who will masquerade as God, being a False Savior), these comets and detritus of Phoenix will rain upon the kingdoms of men and thereafter be known as the Wrath of God. This will be a time of unprecedented quake and volcanic activity, plagues, chemical contamination from space, and a global epidemic of starvation and chaos while meteorites continually strike the earth, vaporizing entire cities and scattering the people of earth into clans and tribes of a dark Neolithic Earth.

Fortunately, most of us will not be alive when this occurs. This is still a long way off into the future. But 2040 AD is *not*. . .

XII

Conclusion of the Phoenix Thesis

The following facts are the core conclusions made by this author in this work:

1. Every 138 years planet Phoenix passes through the inner solar system between Earth and the sun on a north-to-south trajectory.

2. The 138 year orbit was known to the ancients and transmitted to us through legends and olden texts.

3. As 138 years is 1656 months and the Pre-Flood world's history was 1656 years to its ruin at the Great Flood, there seems to be some intelligent design involved in this astronomical cycle; the Age of the Phoenix is 3312 years (1656 + 1656 years from 3895-583 BC).

4. The Phoenix orbits pinpoint 3895 BC as Year One of the 6000 year timeline of Man's curse.

5. The Cursed Earth system is predicated upon the fact of three Phoenix orbits of 138 years each for 414 years.

6. The Phoenix Cycle is predicated upon the fact of four Phoenix orbits of 138 years each for 552 years.

7. The Age of the Phoenix so popular in antiquity (Suns of the ancients) was due to the 552 year periodicity of the sun darkening by Phoenix transits at 2239 BC, 1687 BC, 1135 BC and 583 BC, the Phoenix Legend falling into disrepute because Earth fell out of synch with Phoenix.

8. When Phoenix directly transits between Earth and the sun, the sun darkens and the moon appears as red as blood.

9. Planet Phoenix is covered in a frozen liquid surface that is constantly fracturing and breaking, sending immense cosmic dust clouds careening toward earth along with comet groups and asteroid trains.

10. Planet Phoenix is currently realigning for a direct transit between Earth and the sun, first noticed in 1764 AD by Hoffman and reconfirmed in 1902 AD as Earth was bathed in a dense cosmic dust cloud.

11. Planet Phoenix is scheduled to return to the inner system in 2040 AD and will be in direct transit; the sun will darken and the moon will

appear as blood, global quakes will ruin entire civilizations and cities will vanish, an event foretold in the prophetic annals from the biblical records and ancient texts from around the world.

12.   This work merely scratched the surface to an edifice we have been studying for centuries. We usually only find what we are most diligently searching for, therefore many other Phoenix-related gemstones have been lost to our prejudicial eyes.

What this author has avoided in this book is the *origin* of planet Phoenix, and for that matter, Earth itself. We are today living in a post-cataclysmic era of a former *binary* star system. This sister companion is now a compressed star, a Dark Star known to our predecessors as the Black Sun. This history and accommodating evidence, especially of the Pre-Adamic World and its ruin, is addressed more fully in *Anunnaki Homeworld*, for Phoenix and NIBIRU both orbit our sun along the Dark Star's ecliptic.

These orbital histories maintain a mathematical precision far superior to what is today considered scientifically acceptable, and the following facts demonstrate just how incredible these histories are:

a.   Thousands of libraries throughout history with millions of texts have been accidentally or deliberately destroyed and we no longer have these accounts. This is evident from fact that Thales, Aristarchus, Anaxagors, Lucretius, Pliny and others provided detailed information about Phoenix they had gathered from the books available in their time;

b.   Many witnesses throughout antiquity were either illiterate or their writings were transmitted on perishable media (paper, leaves, bones, bark, wood tablets, parchment, skins, etc.);

c.   Inclement weather, thick overcast or even daylight would have prevented anyone from seeing and recording the transit of Phoenix, or for that matter, *any* astronomical phenomena;

d.   Many appearances of Phoenix, comets or asteroids that would ordinarily affect Earth in a direct transit go unnoticed because these celestial objects pass over the ecliptic on the other side of the sun, or too far to be seen.

Because the author did not desire to disrupt the ebb and flow of his work with data that is readily available in other books, he has provided abbreviated appendices on such topics as are related to Phoenix, but are mentioned in the often vague and undated stories of peoples long past.

Appendix A

Effect of 2040 AD Phoenix Transit

The world we know and live upon is merely a thin skin 30-40 miles thick called the lithosphere. When Phoenix transits it will be closer to Earth than the sun, its presence altering the heat gradient underneath the lithosphere in the nether regions of our planet where the rock is already so hot it remains plastic and viscous. The proximity of Phoenix will create a powerful gravitational affect on our world. At present the lithosphere is attached and in motion with the lower asthenosphere and mantle, however, once the heat builds, radioactivity increases, tidal stress and electromagnetic friction between these two planets will cause the lithosphere to *float*. The crust of the planet will *wander* and as the prophetic texts like the Revelation indicate, the sun will go down at noon, it will darken in the sky, the fixed stars shall appear in daytime and then fall rapidly over the horizon and earthquakes will shake terribly the earth. Commonly called a pole shift, earth will virtually list over and roll as Phoenix weakens Earth's gravitational link to the sun.

The Great Pyramid site at 30 degrees north of the equator will be the epicenter of motion, the Giza plateau merely spinning very slowly as regions to the west are shoved southward and the far east is pushed northward. North America will move south into the tropics and area of present Central America while South America will be thrust into the Antarctic. Those lands passing over the equatorial bulge will be the worst afflicted. The Arctic Circle will descend to the position of present day United States and far to the south the Antarctic will be shoved northward to the region of present day Australia while Australia takes the place of upper Indonesia and China. China and the East will be moved north to the area of Russia and Siberia, while Siberia will become the new Arctic. Crustal plates will collide, buckle, fracture and slide under one another, laying waste to their surface topography.

Many cities of the world, including New York, will be swept out to sea. Entire hinterlands will lower into the depths of the ocean and multitudes of islands around the world will be lost as others appear. The closer a nation is located to Israel and Egypt the less destruction it will suffer. Tsunamis will pull all surface materials and millions of people out to sea and volcanic resurfacing will bury alive more people than 10,000 Vesuvius's. Worldwide time-sensitive power grids will go permanently offline and all systems relying on satellite technology will be rendered useless as everything orbiting earth, including the moon, is steadily bombarded by detritus in the tail of Phoenix as it fragments near the Earth. Glacial-sized meteorites will melt entering

Earth's atmosphere and gravel-to-boulder size rocks will plow through all orbiting stations and satellites as electromagnetic storms attended by gale-force winds will blanket our world. Vessels aloft and out at sea will never be seen again.

Despite this horrific scenario, this 2040 AD event only precedes much more cataclysmic ruin of our planet scheduled in 2046 AD, which is the subject of *Anunnaki Homeworld*.

Appendix B

Legends and Mythos of the Sun Darkening

This information is provided as an Appendix because there is nothing novel about such stories. That the sun darkened in antiquity is well attested widely and documented by many modern authors.

The oldest accounts of the sun darkening that do not refer to an ordinary eclipse come down to us from Babylonia, Egypt, India and China. A Babylonian fragment describes the oncoming of the deluge, reading that ". . .as soon as something of dawn showed in the sky, a *black cloud* from the foundation of heaven came up. . . every gleam of light was turned into darkness. . . a whole day long. . . brother saw not brother." (1) Zecharia Sitchin provides two excellent fragments. An undated Sumerian record reads, "On that day, when heaven was crushed, and the Earth was smitten, its face obliterated by the maelstrom – when the *skies were darkened* and covered as with a *shadow*. . ." (2) He also published this:

> *When the sun reaches its zenith and is* dark,
> *the unrighteous of the land will come to*
> *nought.*
>
> —Prophecy Text of Babylon (3)

The Egyptians convey that in remote times the sun was injured severely and the moon was entirely swallowed by a god. This evil deity was slain and the sun appeared as the moon was vomited back up. (4) These events refer to the Flood and probably the 1687 BC sun darkening, a date relatively contemporary to this Vedic text from the *Mahabharata* of India:

> Dense arrows of flame, like a great shower,
> issued forth upon creation, encompassing the
> enemy. . . a thick gloom swiftly settled upon
> the Pandeva hosts. All points of the compass
> were *lost in darkness*. Fierce winds began
> to blow. Clouds roared upward, showering
> dust and gravel. . . the earth shook. . . from
> all points of the compass the arrows of
> flame rained continuously and fiercely. (5)

111

The Chinese preserved a tradition saying that the sun went dark during the reign of Emperor Kung Change whose official astrologers failed to predict it. They were decapitated, for they neglected to shoot their arrows at the sun, as was their traditions to help the sun god begin his light. (6) The Japanese too hold that long ago the sun hid in a cave and the land was covered in blackness. (7) Older Arabic traditions concerning Adam before the Flood mention that at his death the sun darkened, a fact attested to in *Anunnaki Homeworld*. (8) Even Ovid wrote that ". . .a day went without the sun." (9)

Even on the other side of the planet the stories were passed down from the distant past. High in the Andes mountains the native Americans claim that in a disaster long ago the sun went out for five days. (10) The sun darkening subject is a constant theme in the songs of Native Americans, according to Spanish chronicler and writer Antonio Herrera (1549-1625 AD). (11) The ancient Mexicans believed that when Quetzalcoatl vanished long ago, the sun and moon darkened and only a single star was visible in the sky. (12) This is consistent with the 33 AD event caused by the Vials of Phoenix Comet Group. To the far north in Alaska the Tanaina people recall a time when a rich man stole the sun and moon when giants walked the earth. (13)

There are too many traditions concerning the sun going dark in antiquity to mention, and the majority of them in North America all have in common that someone caught the sun in a trap, and remarkably, these stories all link the event with the massive stone building projects of the Megalithic Age, (14) the period from the 1687-1135 BC sun darkenings.

Appendix C

When the Sun Stood Still

The biblical account of the sun standing still remarkably defers to the authority of the *Book of Jasher*, asserting, ". . .is this not written in the Book of Jasher?" The event was unprecedented, and has not occurred since. As found in this book the year the sun stood still (Earth ceased rotating while gravitationally locked to a near-colliding gigantic object) was in 1407 BC.

In the *Apocalypse of Baruch* (Baruch was a scribe of Jeremiah the prophet), we read that ". . .the heavens at that time were shaken from their place," (1) and in the Old Testament book of Ecclesiastes there is a record stating that ". . .was not one day as long as two?" (2) Flavius Josephus wrote that ". . .he (Joshua) understood that God assisted him, which he declared by thunder and thunderbolts, as also by the falling of hail larger than unusual. Moreover it happened that the day was *lengthened*. . . . Now, that the day was lengthened at this time, and was longer than ordinary, is expressed in the books laid up in the Temple." (3) One of these books was the Temple copy of the *Book of Jasher*.

Persian writings mention that long ago a single day became as long as three. (4) Chinese records mention that the sun did not set for a long time and the land was burned. (5) Tibetan tales tell of a demigod named Milarepa who was an undesirable being who caused the sun to cease going down in the sky in the early evening, and made it hang over the edge for a very long time. (6) On the other side of the world Aztec Nahautl stories told of a time long ago when the *sun failed to rise* for a long time. (7)

To learn more of this solaric anomaly refer to the works of Immanuel Velikovsky, Zecharia Sitchin and Charles Hapgood.

Bibliography of Cited Works

The Dragon Legacy: Secret History of an Ancient Bloodline: Nicholas de Vere (Book Tree)

The Wars of Gods and Men: Zecharia Sitchin (Avon)

Mystery at Acambaro: Charles Hapgood, 1973 (Adventures Unlimited)

The Greek Myths: Robert Graves (Penguin)

Tracing Our Ancestors: Frederick Haberman (Covenant Publishing, London)

The Book of Jasher: (Book Tree)

Natural History: Pliny (Penguin) Trans. John F. Healy

On the Nature of the Universe: Ronald Melville (Oxford World's Classics) Lucretius

The Great Pyramid: Its Divine Message: D. Davidson and H. Aldersmith, 12th Edition, 1924 (Book Tree reprint)

The Histories: Herodotus, trans. Aubrey de Selincourt (Penguin)

Philosophy of Aristotle: Renford Bambrough, 1963 (Mentor)

The Destruction of Atlantis: Frank Joseph, 2002 (Bear & Co.)

Atlantis: Mother of Empires: Robert B. Stacey-Judd, 1939 (Adventures Unlimited)

The Stone Angle: Jack Wun

Celtic Myth and Legend: An A-Z of People and Places: Mike Dixon-Kennedy (Blandford)

Nature Worship: Ted St. Rain, 1999 (Book Tree)

When Men are Gods: G. Cope Schellhorn (Horus House Press)

Our Haunted Planet: John Keel; (Glade Press) 2002, originally published 1971

Egypt, Greece and Rome: Charles Freeman (Oxford University Press)

The Shadow of Atlantis: Alexander Braghine (Adventures Unlimited) 1940

Lost Cities and Ancient Mysteries of South America: David Hatcher Childress (Adventures Unlimited)

Our Cosmic Ancestors: Maurice Chatelain (Temple of Golden Publications)

Atlantic in America: Lewis Spence, 1925 (Book Tree reprint)

Riddle of the Pacific: John Macmillan Brown, 1924 (Adventures Unlimited)

Voyages of the Pyramid Builders: Robert Schoch and Robert McNally (Tarcher Putnam)

Secret Cities of Old South America: Harold T. Wilkins, 1952 (Adventures Unlimited)

Lost Cities of China, Central Asia and India: David Hatcher Childress (Adventures Unlimited)

USA Today Report: Earlier Mount Vesuvius Blast Should be Warning to Naples, by Dan Vergano, 2006

The Natural Genesis: Gerald Massey, 1883 (Black Classic Press)

The Search of Noah's Ark: David Balsiger & Charles E. Sellier, Jr. (Sun Classic Books)

The Merovingians Mythos: The Mystery of Rennes-le-Chateau, Tracy R. Twyman (Dragon Key Press)

The Archko Volume: Drs. McIntosh & Twyman (Book Tree)

The Lost Books of the Bible & Forgotten Books of Eden: (World Bible Publishers)

Talmudic Sanhedrin: Rashi, 1050 CE, Leviticus Rabbi

Cataclysm! Compelling Evidence of a Cosmic Catastrophe in 9500 BC, D.S. Allen J.B. Delair (Bear & Co.)

Nostradamus: The Complete Prophecies: Mario Reading. 2006 (Watkins Publishing, London)

The Lost Realms: Zecharia Sitchin (Avon)

The Secrets of Time: Stephen Jones (God's Kingdom Ministries)

The Round Towers of Atlantis: Henry O'Brien, 1834 (Adventures Unlimited)

The Gods of Eden: William Bramley (Avon)

Sages and Seers: Manly P. Hall (Philosophical Research Society)

Today, Tomorrow and the Great Beyond: John S. Fox (Assn. of Covenant People)

Antiquities of the Jews: Flavius Josephus, trans. 1736 by William Whiston (Hendrickson Pub.)

Biblical Antiquities III: E. Raymond Capt (Artisan)

Great Disasters: Readers Digest Assoc. edited by Kaari Ward, 1989

Secret Places of the Lion: George Hunt Williamson (Destiny)

The End of Days: Zecharia Sitchin, 2006 (William Morrow)

A Short History of the World: H.G. Wells (Book Tree)

National Sunday Laws: A. Jan Marcussen (Amazing Truth Publications)

Invisible Residents: Ivan T. Sanderson, 1970 (Adventures Unlimited)

Atlantis: The Antediluvian World: Ignatius Donnelly & Egerton Sykes (reprinted by Book Tree)

2007 Almanac: (World Almanac Publisher)

Book of the Damned: Charles Fort, 1919 (Book Tree)

The Discoverers: A History of Man's search to know His World and Himself: Daniel J. Boorstin, 1983 (Vintage)

A Sorrow in Our Heart: Allen W. Eckert (Bantam)

Three Books of Occult Philosophy: Henry Cornelius Agrippa, annotated by Donald Tyson (Lewellyn), 16th century

Psychic and Occult Views and Reviews Magazine: (Psychic Review Company) St. Clair, St. Toledo, OH, Martha J. Keller Secretary

Secrets of Nostradamus: David Ovason (Harper Collins)

Prophecy Flash: William Dankenbring, Vol. 21 No. 6, Jan-Feb. 2008

Journeys to the Mythical Past: Zechairah Sitchin 2007 (Bear & Co.)

America's Secret Destiny: Spiritual Vision of the Founding of a Nation: Robert Hieronimus, PH D (Destiny)

When Time Began: Zecharia Sitchin (Avon)

Twilight of the Idols: Frederick Nietzsche (Walter Kaufmann: Vintage)

The Histories: Tacitus, trans. Alfred church & William Brodribb (Penguin Classics)

Dialogues of Plato: Jowett Translational edited J.D. Kaplan (Pocket Library, 1955)

Origin and Evolution of Freemasonry: Albert Churchward, 1920 (Book Tree)

The Great Secret: Life's Meaning as Revealed Through Ancient, Hidden Traditions: Maurice Maeterlinck, 1922 (Book Tree)

Beyond Good and Evil: Frederick Nietzsche (Walter Kaufmann: Vintage)

A Study of Numbers: R.A. Schwaller de Lubicz, 1950 (Inner Traditions)

Technics and Civilization: Lewis Mumford (Harcourt, Brace & World, Inc.)

Uncommon Sense: The Real American Manifesto: William James Murray (America West Publishers)

Symbols, Sex and the Stars: Earnest Busenbark (Book Tree)

Light of Egypt: The Science of the Soul and Stars: Vols. I and II, Thomas Burgoyne (Book Tree)

Doctrine of Sin in Babylonian Religion: Julian Morganstern (Book Tree)

The Book of Enoch: (Artisan; also Book Tree)

The Later Roman Empire: Ammianus Marcellinus, trans. Walter Hamilton (Penguin)

The First Genesis: William Dankenbring (Triumph Publishing)

The Early History of Rome: Livy, trans. Aubrey de Selincourt (Penguin)

Annals of Imperial Rome: Tacitus, trans. Michael Grant (Penguin)

Poleshift: John White (ARE Press)

Aryan Sun Myths: The Origins of Religions: Sarah Elizabeth Titcomb, 1899 (Book Tree)

Pyramid Quest: Robert Schock and Robert Aquinas McNally (Tarcher Penguin)

1066 AD The Year of the Conquest: David Howarth (Barnes & Noble)

Prophets of Doom: Daniel Cohen (The Millbrook Press)

Commentaries on Occult Philosophy of Agrippa: Willy Schrodter (Samuel Weiser)

End of Eden: The Comet that Changed Civilization: Graham Phillips (Bear & Co.) 2007

Far Out Adventures: edited by David Hatcher Childress (Adventures Unlimited)

Tales of the Patriarchs: Great Books of the Islamic World: Muhammed ibn abd Allah al-Kisai, trans. W.M. Thackston Jr. (Great Books of the Islamic World)

Northern Tales: Howard Norman (Pantheon)

Notes and References

I. Existence of Planet Phoenix

1. *The Dragon Legacy* 210, 212
2. *The Wars of Gods and Men* 39
3. *Mystery at Acambaro*: intro. 34
4. *The Greek Myths* 196
5. *Tracing Our Ancestors* 33
6. *Tracing Our Ancestors* 33
7. *Book of Jasher*
8. *Tracing Our Ancestors* 22
9. *Tracing Our Ancestors* 22
10. *Natural History: Universe and the World* 97 p. 22
11. *Natural History: Man* 7 p. 75
12. *Natural History: Universe and the World* 147 p. 28
13. *On the Nature of the Universe*: Book 5, lines 324-330; Book 6 lines 290-295; 585-590
14. *On the Nature of the Universe*: Book 5 lines 747-768
15. *The Great Pyramid: Its Divine Message* 100
16. *The Great Pyramid: Its Divine Message* 103
17. *Natural History: Universe and the World* 53 p. 19-20

II. Age of the Phoenix and Cycle of Cataclysm

1. *Histories, Book VII* 152 p. 421
2. *Histories, Book I* 74 p. 30
3. *Philosophy of Aristotle* 348; *Ethics VI* p. 419; *Poetics*
4. *Destruction of Atlantis* 169
5. *Atlantis: Mother of Empires* 237
6. *The Stone Angle* 46
7. *Celtic Myth and Legend* 108-109
8. *Book of Jasher* 10:23
9. *Nature Worship* 19
10. *Book of Jasher* 37:17-18
11. *When Men are Gods* 177-178
12. *Our Haunted Planet* 32
13. *Egypt, Greece and Rome* 78-79
14. *The Shadow of Atlantis* 201; *Lost Cities and Ancient Mysteries of South America* 139
15. *Lost Cities and Ancient Mysteries of South America* 139
16. *Our Cosmic Ancestors* 159

17.   *Lost Cities and Ancient Mysteries of South America* 138; *Atlantis in America* 258
18.   *Atlantis: Mother of Empires* 200
19.   *Riddle of the Pacific* 15
20.   *Riddle of the Pacific* 15
21.   *Riddle of the Pacific* 150
22.   *Voyages of the Pyramid Builders* 197-198
23.   *The Great Pyramid: Its Divine Message*: Appendix
24.   *Secret Cities of Old South America* 384
25.   *Lost Cities of China, Central Asia and India* 361
26.   *Our Haunted Planet* 54-55
27.   *USA Today Report: Earlier Mount Vesuvius Blast Should Be Warning to Naples*, by Dan Vergano, 2006
28.   *The Great Pyramid: Its Divine Message* 5, 52
29.   *The Natural Genesis Vol. II* p. 241
30.   *The Destruction of Atlantis* 140
31.   *The Book of Jasher* 6:11, 13
32.   *The Flood Reconsidered* 7, cited in *In Search of Noah's Ark* 61
33.   *In Search of Noah's Ark* 61
34.   *The Merovingian Mythos* 182-183

III.   Planet of Two Calendars

1.   *The Archko Volume*
2.   *Genesis* 3:14-24
3.   *The Great Pyramid: Its Divine Message* 105, 76-77
4.   *The Great Pyramid: Its Divine Message* 555
5.   *Book of Adam and Eve* II 19:1, 20:15
6.   *Rashi, Talmudic-Sanhedrin* 97A 1050 CE, *Leviticus Rabbi* 29:1
7.   *Secret Cities of Old South America* 391
8.   *Cataclysm!* 12
9.   *The Merovingian Mythos* 186, note 25

IV.   Phoenix Cycles Demonstrated

1.   *Nostradamus: The Complete Prophecies* 265
2.   *The Lost Realms*: Sitchin; *Strangers From Across the Sea*
3.   *Daniel* 8:9-14
4.   *Egypt, Greece and Rome* 607
5.   *Secrets of Time* 207
6.   *Round Towers of Atlantis* 60
7.   *The Dragon Legacy* 212
8.   *The Gods of Eden: Bramley* 181-183
9.   *Sages and Seers* 10

10.   *Today, Tomorrow and the Great Beyond* 181-182
11.   *Biblical Antiquities III* 41
12.   *Great Disasters* 175
13.   *Great Disasters* 176
14.   *Secret Places of the Lion* 91

V.   Cursed Earth Periods

1.   *Genesis* 9:25-27
2.   *Genesis* 21:22-34, *Jasher* 22:1-10
3.   *Exodus* 17:8-16
4.   *Exodus* 17:14-16, *1 Sam.* 17:8-9, 32-33
5.   *1 Samuel* 28:7, 31:6
6.   *2 Esdras* 2:13, *2 Kings* 24:13
7.   *1 Kings* 6:1
8.   *1 Kings* 6:1, 37-38
9.   *Genesis* 15, *Book of Jasher* 13:17-19
10.   *Jasher* LXXV 1-18, *Joshua* 14:7, *1 Chronicles* 7:20-22
11.   *Secrets of Time* 207
12.   *End of Days* 200-201
13.   *The Lost Realms*; also *The Day the Sun Stood Still*
14.   *The Great Pyramid: Its Divine Message* 323-326 A. Tables XVII-XIX

VI.   Modern Cursed Earth Periods

1.   *Great Disasters* 47
2.   *A Short History of the World* 132
3.   *National Sunday Laws*: A. Jan Marcussen (Amazing Truth Publications)
4.   *Atlantis in America* 204 (modern version, Book Tree)
5.   *The Dragon Legacy* 89, 143
6.   *The Dragon Legacy* 89
7.   *Invisible Residents* 32
8.   *Voyages of the Pyramid Builders* 211
9.   *Sages and Seers* 11-12
10.   *Cataclysm!* 349
11.   *Destruction of Atlantis* 173; *Atlantis: The Antediluvian World* 35
12.   *2007 Almanac*, p. 270

VII.   Orbital Chronology of Planet Phoenix

1.   *Cataclysm!* 350; *Book of the Damned* 149
2.   *Book of the Damned* 146-147

3.   *Cataclysm!* 349
4.   *Book of the Damned* 170
5.   *The Discoverers* 586
6.   *A Sorrow in Our Heart* 39-41
7.   *Three Books of Occult Philosophy* 161
8.   *The Dragon Legacy* 212
9.   *Psychic and Occult Views and Reviews* 74-75
10.   *Secrets of Nostradamus* 307
11.   *Cataclysm!* 200
12.   *Atlantis: The Antediluvian World* 37; *Great Disasters* 164
13.   *Prophecy Flash,* Vol. 21 No. 6
14.   *Book of the Damned* 24-26, 28
15.   *Book of the Damned* 144
16.   *Book of the Damned* 170
17.   *Book of the Damned* 25
18.   *2007 Almanac,* p. 270
19.   *Book of the Damned* 172
20.   *Book of the Damned* 171
21.   *Journeys to the Mythical Past* 215

VIII.   Secret Calendar of the Great Seal of the United States of America

1.   *America's Secret Destiny* 44-45, 52, 85
2.   *When Time Began* 55
3.   *The Wars of Gods and Men* 172
4.   *The Great Pyramid: Its Divine Message* 172
5.   *2 Esdras* 14:45
6.   *Twilight of the Idols* 61
7.   Tacitus, *The Histories* 142
8.   Lucretius, *On the Nature of the Universe, Book IV* 474-480
9.   *Natural History: Man* 7 p. 75
10.   *Dialogues of Plato* 213, 90
11.   *Dialogues of Plato* 73
12.   *Origin and Evolution of Freemasonry* 209
13.   *The Great Secret* 9
14.   *Sirach* 38:39, 38:32
15.   *Beyond Good and Evil* 35
16.   *Philosophy of Aristotle* 353, *Ethics Book III*
17.   *A Study of Numbers* 9
18.   Lucretius, *On the Nature of the Universe, Book IV* 812-817
19.   *Philosophy of Aristotle* 217

IX.   2040 AD Return of Planet Phoenix

1.   *Isaiah* 46:9-11
2.   *Technics and Civilization* 16
3.   *Uncommon Sense* 27
4.   *Dialogues of Plato* 12-13, citing Socrates
5.   Herodotus, *Histories, Book III* 155 p. 214
6.   *Symbols, Sex and the Stars* 270
7.   *Light of Egypt Vol. II* p. 210
8.   *Light of Egypt Vol. II* 168
9.   *Doctrine of Sin in the Babylonian Religion* 14-15
10.   *Book of Enoch* 55:4
11.   *Psalm* 46:2-3
12.   *Isaiah* 13:10-13
13.   *Isaiah* 24:1
14.   *Ezekiel* 32:7-8
15.   *Joel* 2:10, 32
16.   *Book of the Damned* 48
17.   *Three Books of Occult Philosophy* 673

X.   The Joshua Comet Group

1.   *Cataclysm!* 351
2.   *Ammianus Marcellinus, Book 22*, p. 254
3.   *Three Books of Occult Philosophy* 675, Note 31; Pliny, *Natural History: Universe and the World* 149 p. 29; *The First Genesis* 192
4.   *The Destruction of Atlantis* 169
5.   *Secrets of Time* 208
6.   *Early History of Rome*; *Livy* 3.5
7.   *Livy* 3.10
8.   *Livy* 4.21
9.   *Livy* 5:12-13
10.   Pliny, *Natural History: Universe and the World* 96 p. 21
11.   *Archko Volume* 139
12.   Tacitus, *Annals VI*, 25-28, p. 213-214
13.   *Gospel of Nicodemus* 8:1
14.   *Archko Volume* 141
15.   *Lost Books of the Bible* 274-275
16.   *Lost Books of the Bible* 274-275
17.   *Archko Volume* 142
18.   *Lost Books of the Bible* 277
19.   *Archko Volume* 37
20.   *Poleshift* 176
21.   *The Lost Realms*; *The Day the Sun Stood Still*

22.   *Aryan Sun Myths*: 38-40
23.   *Sages and Seers* 11

XI.   Vials of Phoenix Comet Group

1.   *The Discoverers* 75
2.   *Pyramid Quest* 238
3.   *Round Towers of Atlantis* 60
4.   *Atlantis in America* 50
5.   *Round Towers of Atlantis* 50
6.   *1066 AD The Year of the Conquest* 83
7.   *Voyages of the Pyramid Builders* 210
8.   *Gods of Eden*: Bramley 180-185, *Sages and Seers* 9-12
9.   *Atlantis in America*: Spence 175
10.   *Lost Cities and Ancient Mysteries of South America* 56-57
11.   *Prophets of Doom* 122
12.   *Sages and Seers* 19
13.   *Commentaries of Occult Philosophy of Agrippa* 67
14.   *Book of the Damned* 37, 41-43, 49
15.   *Commentaries on Occult Philosophy of Agrippa* 67-68
16.   *Book of the Damned* 47
17.   *End of Eden* 164
18.   *Cataclysm!* 200
19.   *Book of the Damned* 15
20.   *Book of the Damned* 162; *Cataclysm!* 349
21.   *Book of the Damned* 179
22.   *Book of the Damned* 48, 55
23.   *Book of the Damned* 129
24.   *Book of the Damned* 48
25.   *Prophets of Doom* 30
26.   *Book of the Damned* 37, 211
27.   *Book of the Damned* 162, 200
28.   *Cataclysm!* 200
29.   *Book of the Damned* 36
30.   *Book of the Damned* 44, 184
31.   *Book of the Damned* 148-150
32.   *Book of the Damned* 154
33.   *Far Out Adventures* 375
34.   *Book of the Damned* 169
35.   *Atlantis: The Antediluvian World* 34
36.   *Book of the Damned* 21, 39, 137, 147
37.   *Book of the Damned* 23
38.   *Biblical Antiquities III* 99
39.   *Secret Cities of Old South America* 39

Appendix B.   Legends and Myths of the Sun Darkening

1.   *Atlantis: The Antediluvian World* 63
2.   *When Time Began;* chapter 12 – Age of the Ram
3.   *When Time Began,* chapter 13 – Aftermath
4.   *Book of the Damned* 343
5.   *Lost Cities of Ancient Lemuria and the Pacific* 73
6.   *Our Cosmic Ancestors* 108
7.   *Sun Lore of All Ages* 206
8.   *Tales of the Patriarchs* 84
9.   *The Destruction of Atlantis* 117
10. *Secret Cities of Old South America* 48
11. *Secret Cities of Old South America* 49
12. *Secret Cities of Old South America* 100
13. *Northern Tales* 38, 51-52
14. *Sun Lore of All Ages* 212

Appendix C.   When the Sun Stood Still

1.   *Apocalypse of Baruch* LIX:3
2.   *Ecclesiastes* XLVI:4
3.   *Antiquities of the Jews* 5.1.17
4.   *Poleshift* 117, citing Velikovsky
5.   *Poleshift* 117, citing Velikovsky
6.   *Lost Cities of China, Central Asia and India* 64
7.   *Lost Cities of North and Central America* 255

About the Author

As of 2009 Jason M. Breshears has been in a south Texas prison for over 19 years, since he was 17 years old. He was given an agreed-to sentence that would require him to serve only seven and a half years in prison. In 1999 he was granted his parole release but Texas Parole Board adopted new retroactive policies that have since blocked his release from prison. Though a model prisoner and published author, Jason has been denied parole release five times and been made to serve over twelve years more than what his original plea bargain with the State mandated. His situation is not an anomaly in the draconian system of Texas politics. Until he is released he continues his research and writing, and at 36 years old has written the following works:

Lost Scriptures of Giza: Enochian Mysteries of the World's Oldest Texts
When the Sun Darkens: Orbital History and 2040 AD Return of Planet
    Phoenix
Anunnaki Homeworld: Orbital History and 2046 AD Return of Planet
    NIBIRU
Descent of the Seven Kings: Anunnaki Chronology and 2052 AD Return
    of the Fallen Ones
Chronotecture: Lost Science of Prophetic Engineering
Chronicon: Timelines of the Ancient Future
King of the Giants: Mighty Hunter of World Mythology
The Book of Jason: Philosophical Musings of a Dark Prophet

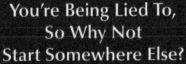

Printed in June 2023
by Rotomail Italia S.p.A., Vignate (MI) - Italy